Lecture Notes
in Physics

Edited by J. Ehlers, München, K. Hepp, Zürich, and
H. A. Weidenmüller, Heidelberg

32

Particles, Quantum Fields and Statistical Mechanics

Springer-Verlag
Berlin · Heidelberg · New York

Lecture Notes in Physics

Edited by J. Ehlers, München, K. Hepp, Zürich, and
H. A. Weidenmüller, Heidelberg
Managing Editor: W. Beiglböck, Heidelberg

32

Particles, Quantum Fields and Statistical Mechanics

Proceedings of the 1973 Summer Institute in Theoretical Physics held at the Centro de Investigacion y de Estudios Avanzados del IPN – Mexico City

Edited by M. Alexanian and A. Zepeda

Springer-Verlag
Berlin · Heidelberg · New York 1975

Editors:
Prof. Dr. M. Alexanian
Prof. Dr. A. Zepeda
Centro de Investigacion del IPN
Departamento de Fisica
Apartado Postal 14-740
Mexico 14, D.F., Mexico

Library of Congress Cataloging in Publication Data

Summer Institute in Theoretical Physics, Centro de Investigación y de Estudios Avanzados del IPN, 1973.
Proceedings of the 1973 Summer Institute in Theoretical Physics held at the Centro de Investigación y de Estudios Avanzados del IPN, Mexico City.

(Lecture notes in physics ; v. 32)
Bibliography: p.
Includes index.
CONTENTS: Blankenbecler, R. Large momentum transfer scattering and hardonic bremsstrahlung.--Symanzik, K. Small-distance behaviour in field theory.--Alder, B. J. Computations in statistical mechanics.--Frishman, Y. Quark trapping in a model field theory.
1. Particles (Nuclear physics)--Congresses. 2. Field theory (Physics)--Congresses. 3. Statistical mechanics--Congresses. I. Alexanian, M., 1936- ed. II. Zepeda, Arnulfo, 1943- ed. III. Series.
QC793.S85 1973 539.7'21 74-28357

ISBN 3-540-07022-2 Springer-Verlag Berlin · Heidelberg · New York
ISBN 0-387-07022-2 Springer-Verlag New York · Heidelberg · Berlin

Offsetprinting and bookbinding: Julius Beltz, Hemsbach/Bergstr.

FOREWORD

The Centro de Investigación has been holding yearly summer schools in theoretical physics since 1969.

This first publication is respectfully dedicated to the memory of Dr. Arturo Rosenbluth co-founder and first director of the Centro. Dr. Rosenbluth's dedication to high scientific achievements inspired the creation of our summer school.

We would like to thank our present director Dr. Guillermo Massieu for continuing and furthering such scientific endeavors.

The partial financial support of the Fomento Educacional is acknowledged and also the support and interest of Prof. Manuel Sandoval Vallarta.

M. Alexanian

A. Zepeda

CONTENTS

RICHARD BLANKENBECLER
Stanford Linear Accelerator Center, U.S.A.

LARGE MOMENTUM TRANSFER SCATTERING AND HADRONIC BREMSSTRAHLUNG

CONTENTS

RICHARD BLANKENBECLER
SLAC, CALIFORNIA

LARGE MOMENTUM TRANSFER SCATTERING AND HADRONIC BREMSSTRAHLUNG

I. Introduction

Strong interactions are clearly very complicated physical processes and thereby are difficult to understand. If nucleons are composite, however, there is the possibility for finding a kinematic region where the reactions are of a simple nature and perhaps can be readily understood. Hadronic matter seems to be quite heavy stuff and hence the propagation of interactions over a long distance requires that it be arranged in a highly coherent state to take advantage of all possible binding forces and to achieve a small mass. Such states would seem to be very fragile and hence only contribute to small momentum transfer scattering. Large angle scattering on the other hand is determined by the short range interactions and hence one should be able to probe the simplest possible states of hadronic matter which should be the most rugged.

Based on this type of physical argument, one is led to consider a composite model of the hadrons and one expects that at large angles and large energies, the impulse approximation should be valid, and this must be checked in any particular theory that one decides to use. Therefore, the nature and properties of the constituents can be probed and determined and then used to predict other reactions and used to extend these predictions into wider angular ranges.

Generally speaking, composite models of hadrons and their mutual interactions can be divided into two classes. In one type, hadron-hadron interactions are dominated in turn by direct parton-parton interactions. In the other type, parton-parton interactions are "soft" and negligible, except for the effects of binding, and the dominant hadron-hadron force is basically due to the interchange of common constituents. One such theory of the former types which is of considerable popularity and was discussed by Berman *et al* in great detail, uses vector gluons to mediate the parton-parton force. While these theories have several attractive theoretical features, they have considerable difficulty in explaining the huge ratio at large angles of elastic processes such as $(pp)/(\bar{p}p)$ and $(K^+p)/(K^-p)$. The inclusive experiments at large transverse momentum performed at the CERN-ISR also prefer the interchange type of theory. In particular, the reaction $pp \to \pi^0 X$ has an energy dependence at fixed $X_\perp = 2P_\perp/\sqrt{s}$ which does not scale as predicted by the vector gluon theory (s^{-2}), but varies as ($s^{-4\pm0.5}$) which is the interchange prediction with monopole meson form factors as required by the low energy data.

Therefore we will discard models of the first class and discuss only models involving constituent interchange such as discussed by Gunion *et al.* The predictions of any composite model can be made only after one has developed a suitable formalism for describing bound systems. Originally, it was decided to use a formulation using old-fashioned perturbation theory in the infinite momentum frame rather than a covariant, or Bethe-Salpeter, approach such as used by Landshoff and Polkinghorne. Both approaches have advantages and disadvantages. One disavantage of the former is that results are non covariant at intermediate stages of the calculation, but this annoyance is eliminated by using the method of Schmidt. One advantage is that the relativistic bound state wave functions which enter have a structure which is very similar to that of the nonrelativistic case. One can therefore use the insight gained in this familiar situation, since it will be necessary to approximate these wave functions in our explicit calculations. Such insight is missing in the covariant approach. One should take care, however, to separate the basic elements of the interchange model and these practical approximations which are necessary to make specific predictions. In fact, this model will allow a simple description of large angle processes which becomes more complex and involved in a natural way as one moves to smaller angles and hence provides a unified description of both large and small angle scattering with considerable predictive power. The asymptotic form of the scattering amplitudes in the interchange model can be predicted in a manner which is independent of the details of the binding interaction and depends only on the limiting behavior of the wave function. In this sense it provides the most economical and simplest possible description of hadronic processes in the deep scattering region of large s, t, and u. It will be assumed that the hadronic wave functions fall as powers of the momenta rather than, say, exponentially. This will ensure the overall consistency of our impulse approximation expansion and seems to be required by the data. Analytically, this also means that it will be a simple matter to continue to the crossed processes and this will provide a severe test of the theory because each matrix element must correctly predict three different reactions in widely separated regions of the kinematic variables s, t, and u.

The deep scattering region will be discussed first, since the theory is particularly simple there, and then it will be extended into the Regge domain. For a variety of reasons, the calculations are extremely difficult to extend into the region of small momentum transfer, so that we shall have to be content with $|t|>1$ or $2(\mathrm{Gev}/c)^2$ in our quantitative discussion. For smaller $|t|$, we are forced to be qualitative like everyone else.

II. Schmidt Method

Let us now turn to a very convenient method for performing the four dimensional momentum integrals which occur in relativistic theories which was developed by M. Schmidt. Let us first consider the contribution to the form factor illustrated in fig (1 a). The mass of particle a will be denoted by a and similarly for the other particles. It will be assumed that a=b and then the form factor integral is obviously

$$(2p+q)_\mu F(q^2) = ieg^2 \int \frac{d^4k}{(2\pi)^4} (k^2-a^2+i\varepsilon)^{-1} \left((k+q)^2-b^2+i\varepsilon\right)^{-1} \left((p-k)^2-c^2+i\varepsilon\right)^{-1} (2k+q)_\mu$$

The normal procedure is to use Feynman parameters to carry out the k integration. However, following Schmidt, it is more convenient for our purposes to proceed differently. There are two choices we can make at this point. In the infinite momentum method, one parametrizes the four vectors as

$$p = (P + m^2/2P,\ \vec{0}_\perp,\ P)$$
$$q = (q_\perp^2/2P,\ \vec{q}_\perp,\ 0)$$
$$k = \left(xP + (k^2 + k_\perp^2)/2P,\ \vec{k}_\perp,\ xP\right)$$

and then performs the limit $p \to \infty$ after evaluating the invariants in the matrix element. A more satisfactory choice invented by S. Brodsky which does not require that a limit be taken is to write

$$p = (P + m^2/4P,\ \vec{0}_\perp,\ P - m^2/4P)$$
$$q = (q_\perp^2/4P,\ \vec{q}_\perp,\ - q_\perp^2/4P)$$
$$k = \left(xP + (k^2 + k_\perp^2)/4xP,\ \vec{k}_\perp,\ xP - (k^2 + k_\perp^2)/4xP\right)$$

In both the infinite momentum method and in the finite momentum method one computes that

$$d^4k = d^2k_\perp dk^2 dx/2|x|$$

and the limits on k^2 and x are $-\infty$ to $+\infty$.

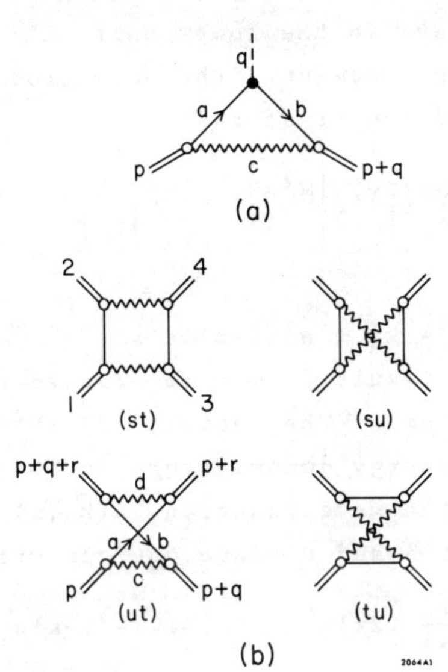
q
a
b
c
p
p+q
(a)
2
4
1
3
(st)
(su)
p+q+r
p+r
d
a
b
c
p
p+q
(ut)
(tu)
(b)
2064A1

FIG. 1

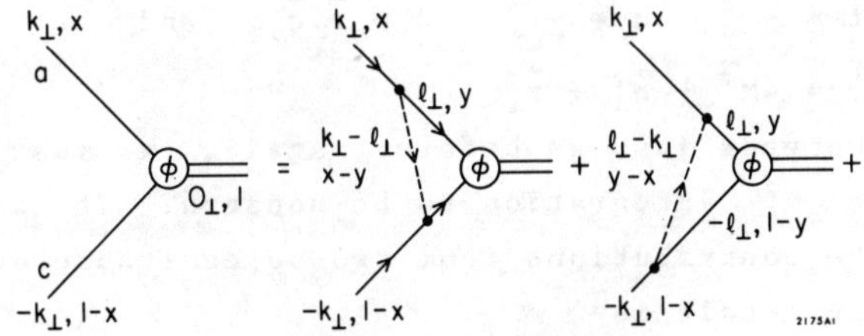
k⊥, x
a
c
-k⊥, l-x
O⊥, l
=
ℓ⊥, y
k⊥- ℓ⊥
x-y
+
ℓ⊥-k⊥
y-x
-ℓ⊥, l-y
+···
2175A1

FIG. 2

Now the integral over k^2 can be carried out directly since only the singularity coming from the pole in the propagator of particle c contributes to the integral. This occurs because the poles from the other two propagators are always in the lower half k^2 plane. For x values between zero and one, however, the pole from particle c is in the upper half plane and one finds that

$$F(q^2) = \frac{1}{2(2\pi)^3}\int d^2k_\perp \int_0^1 \frac{dx\,(2x)}{x^2(1-x)} \left[M^2 - S(\vec{k}_\perp, x)\right]^{-1} \left[M^2 - S\left(\vec{k}_\perp - (1-x)\vec{q}_\perp, x\right)\right]^{-1}$$

where

$$x(1-x)\; S(\vec{k}_\perp, x) = k_\perp^2 + a^2(1-x) + c^2 x$$

This is the same as the result from time-ordered pertubation theory in the infinite momentum frame. The factors of $(M^2-S)^{-1}$ are recognized as the remnants of the energy denominators in this formulation. They are also recognized as the wave function $\psi(\vec{k}_\perp, x)$ of particle M describing its breakup into a and c since one can write

$$F(q^2) = \frac{1}{2(2\pi)^3}\int \frac{d^2k_\perp dx}{x^2(1-x)}\; (2x)\psi(\vec{k}_\perp,\; x)\psi\left(\vec{k}_\perp - (1-x)\vec{q}_\perp,\; x\right)$$

which is very reminiscent of the nonrelativistic expression for the form factor.

The rearrangement graph of (1 b) can be evaluated via the same method. Choosing p, q and the momentum of particle a as before (in the finite frame method for example) and also choosing

$$r = (r_\perp^2/4P,\; \vec{r}_\perp,\; -\, r_\perp^2/4P)$$

where

$$t = -\vec{q}_\perp^{\,2}, \quad u = -\vec{r}_\perp^{\,2}, \quad 0 = \vec{r}_\perp \cdot \vec{q}_\perp, \quad \text{and}$$

$$s = 4M^2 + \vec{q}_\perp^{\,2} + \vec{r}_\perp^{\,2}$$

The calculation proceeds just as before. Again, x must be between zero and one for the k^2 integration to be nonzero. If x is in this range, one gets the contributions from two poles (those due to particles c and d) and the result is

$$M = \frac{1}{2(2\pi)^3}\int \frac{d^2k_\perp dx}{x^2(1-x)^2}\; \psi\;(\vec{k}_\perp, x)\psi\left(\vec{k}_\perp + (1-x)\vec{q}_\perp, x\right)\psi(\vec{k}_\perp - x\vec{r}_\perp, x)$$

$$\times\psi\left(\vec{k}_\perp + (1-x)\vec{q}_\perp - x\vec{r}_\perp, x\right)\Delta(k_\perp, x)$$

where the energy denominator in the numerator Δ is given by

$$\Delta = M^2 - S(\vec{k}_\perp - x\vec{r}_\perp, x) + M^2 - S\left(\vec{k}_\perp + (1-x)\vec{q}_\perp, x\right)$$

This factor can alternatively by interpreted as the transition potential, in which case M takes on a very familiar form. For details of this identification see Section IV of Gunion *et al* (1973) where this formula is derived from a channel Hamiltonian point of view. One can also easily see that M is the sum of the four time orderings shown in Fig. 1b.

III. Wave Functions

A full discussion of the relativistic wave function $\psi(\vec{k}_\perp,x)$ for composite systems has been given and only a brief review will be given here. The wave function $\psi(\vec{k}_\perp,x)$ given in the previous section is analogous to the familiar zero range wave function considered in non-relativistic theory. The factor (M^2-S) can be directly identified with $(E-H_0)$ which is so familiar. It is no surprise then, that the graphs shown in Fig. 2 produce an equation for the wave function of the form

$$\left[M^2-S(\vec{k}_\perp,x)\right]\psi(\vec{k}_\perp,x) = \int d^2\ell_\perp dy V(\vec{k}_\perp,\ell_\perp,x,y)\psi(\vec{\ell}_\perp,y)$$

where V can be easily calculated if desired. The precise form of V will not be necessary, we only need this equation to motivate our primary assumption that in the limits $k_\perp\to\infty, x\sim 0, x\sim 1$, the function behaves as

$$\psi(\vec{k}_\perp,x) \sim N(x)\left[S(\vec{k}_\perp,x)\right]^{-n}$$

where N(x) is a smooth function of x. If this is the case, then it is easy to see that the asymptotic behavior of the form factor is

$$F(q^2) \sim (\vec{q}_\perp^{\,2})^{-n}\int\frac{d^2k_\perp dx}{x(1-x)}\psi(\vec{k}_\perp,x)\left(\frac{x}{1-x}\right)^n N(x)$$

$$\sim (\vec{q}_\perp^{\,2})^{-n}\ln(\vec{q}_\perp^{\,2}/m^2)\; N(o) + \ldots$$

Therefore a knowledge of the fall off of the form factor determines the fall off of the wave function. We shall ignore log factors in the discussions of the formulae from here on.

Now in general, the form factor at zero momentum transfer is (assuming one charged constituent)

$$F(o) = 1 = \int_0^1 dx\, f(x)$$

$$f(x) \equiv \frac{1}{2(2\pi)^3}\int\frac{d^2k_\perp}{x(1-x)}\psi^2(\vec{k}_\perp,x)$$

This function can be interpreted as the fractional longitudinal momentum spectrum of the charged constituent of the particle under consideration. Therefore, it is also directly related to the inelastic structure function and one finds

$$\nu W_2 \equiv F_2(x) = xf(x)$$

where

$$x = 1/\omega = q^2/2m\nu$$

Using the assumed fall off of the wave function ψ, one finds near $x \sim 1$ that

$$\nu W_2 \sim (1-x)^{2n-1}$$

which is the Drell-Yan-West relation between the threshold behavior of νW_2 and the fall off of the form factor.

The above approach can also be extended to a discussion of fixed poles in Compton scattering. One finds that the residue of the fixed pole at $q^2 = 0$ depends on the integral of $f(x)/x$ and for large q^2, the residue falls like the form factor. This very interesting subject is dealt with in a very clear and simple manner in the paper by Close *et al* to which you are referred for details.

IV. Elastic Scattering

Let us now apply the same power law fall off wave function used for the form factor discussion in the previous section to the scattering amplitude. In the case in which particles $A + B \to A + B$, and where we define the form factor index to be the same as the particle label, i.e. $F_I \sim (-t)^{-I}$, one finds in the deep scattering region ($s\sim -t\sim -u >> M^2$) the approximate formula

$$M(u,t) \overset{\sim}{=} sF_A(-s)F_B(t)F_C(u)I(z) + (A\leftrightarrow B)$$

and

$$I(z) \propto \int dx N_A^2 N_B^2 x^{2B-3}(1-x)^{2A-3}\left[(1-x)^2(1-z) + x^2(1+z)\right]^{1-A}$$

The cosine of the center of mass scattering angle is z and in general $I(z)$ is slowly varying around $z \sim 0$. If the x integral is divergent because A and/or B are too small, there may be logarithmic or even stronger modifications of the u and t dependence of $M(u,t)$. If the (st) topology in fig (1 b) contributes to a process, then the amplitude is easily attained by crossing the previous result, and one obtains for large s and t

$$M(s,t) = F_A(-s)F_B(t)\int dx N_A^2 N_B^2 x^{2B-3}(1-x)^{2A-3}\left[-(1-x)^2 t - x^2 s - i\varepsilon\right]^{1-A}$$

The small x region of the integrand will produce an imaginary part to the integral in the general case. However, in all the physical cases that we have considered it is extremely small compared to the real part. The appropriate linear combination of the (u,t) and (s,t) [in our models the (u,s) diagram doesn't enter most reactions] terms depends on the choice of quantum numbers of the constituents. For example, if there are no antiparticles present in A and B, then only the (u,t) graph can contribute. This is the case in proton-proton and proton-neutron scattering. As easily seen by inspection, the general structure of the angular distribution depends upon the relative amount of these two terms which are present. One expects pp and pn elastic scattering to be more or less symmetric around 90° but $\bar{p}p$ should be peaked only in the forward direction. This is in strong contrast to theories with vector gluons, since to lowest order, the angular distributions would be the same.

The energy dependence of the 90° cross section is a simple power law of the form

$$\frac{d\sigma}{dt}(90°) = |M|^2/s^2 \propto s^{-2C}$$

where C is the lesser of 2A + B or 2B + A. In pp scattering, the form factors have a fall off which is fairly close to a dipole for large q^2. Assuming $A = B \sim 2$, one finds

$$\frac{d\sigma}{dt}(pp, 90°) \propto s^{-12}.$$

In the case of pion-nucleon scattering, we have found that a monopole pion form factor fits the data quite well and in this case

$$\frac{d\sigma}{dt}(\pi p, 90°) \propto s^{-8}.$$

A scalar photon would be expected to couple to the constituents with a monopole wave function but when the vector nature of the coupling to the current is taken into account one finds an effective photon form factor of $F_\gamma(t) \sim (-t)^{-1/2}$, and hence using a monopole leads to the prediction

$$\frac{d\sigma}{dt}(\gamma p \to \pi p, 90°) \propto s^{-7}.$$

This is in excellent agreement with the data of Anderson *et al* who find a power fall off of 7.2±0.4.

I shall not describe in detail the inclusion of the effects of spin into the calculations. It turns out that the relation between the energy dependence of scattering and the form factors is unaffected. However the angular dependence is, in general, affected because factors of (u/s) and (t/s) are introduced into M. From now on the results are computed assuming that the constituents have spin one-half and that the proton is a spin one-half particle plus a spin one core. The mesons are bound states of a pair of spin one-half constituents.

A. Nucleon-Nucleon Scattering

The calculation of nucleon-nucleon scattering is quite involved because of spin complications. After much labor, one finds that the spin averaged matrix element has the form

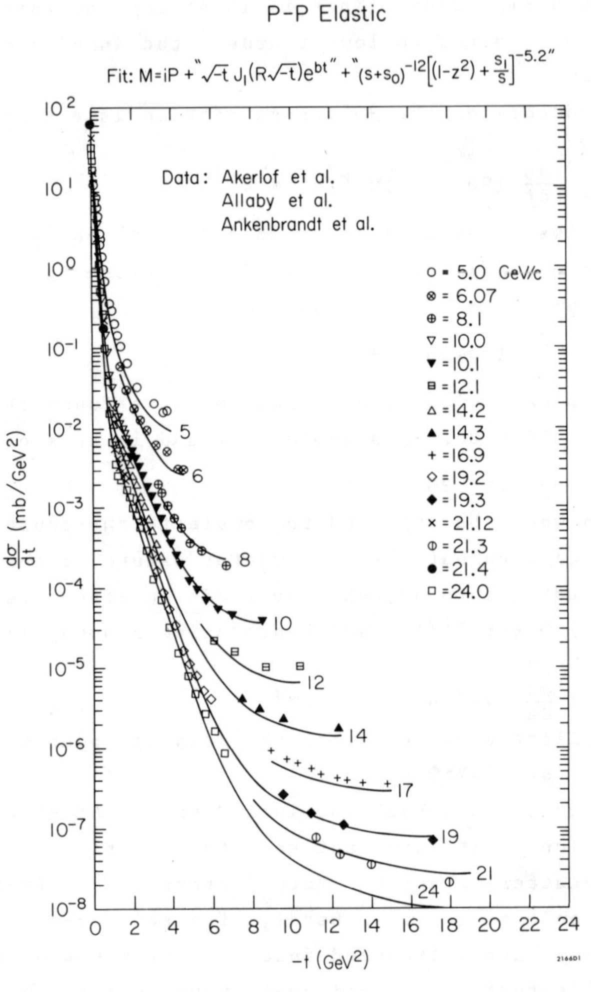
P-P Elastic
Fit: M=iP + "√-t J1(R√-t)e^bt" + "(s+s0)^-12[(1-z²) + s1/s]^-5.2"
Data: Akerlof et al.
Allaby et al.
Ankenbrandt et al.
○ = 5.0 GeV/c
⊗ = 6.07
⊕ = 8.1
▽ = 10.0
▼ = 10.1
⊟ = 12.1
△ = 14.2
▲ = 14.3
+ = 16.9
◇ = 19.2
◆ = 19.3
× = 21.12
⦶ = 21.3
● = 21.4
□ = 24.0
dσ/dt (mb/GeV²)
−t (GeV²)
5
6
8
10
12
14
17
19
21
24
2166D1

FIG. 3

$$M \sim s\,F(-s)F(t)F(u)I(z)$$

and one finds approximately that

$$I(z) \cong (s^2/tu)^{0.6}$$

The predicted cross section for both proton-proton and neutron-proton scattering is of the form

$$\frac{d\sigma}{dt} \cong \sigma_0\, s^{-12}\, (1-z^2)^{-5.2}$$

This prediction is compared with the pp data in Fig 3 in which a pomeron with both a real and imaginary part has been added in to fit the forward direction. When the momentum transfer is larger than a few $(\mathrm{Gev/c})^2$, only the interchange term survives, and one sees that the fit is very good, in spite of the claim by Barger, Halzen and Luthe that it does not fit.

B. Meson-Nucleon Scattering

In general, meson-nucleon scattering has both (u,t) and (s,t) graphs since the mesons must contain anti-constituents in their wave function. The relative weights of these two terms depend on the specific quantum numbers assigned to the basic constituents. The general form for the meson-nucleon differential cross section is

$$\frac{d\sigma}{dt} = \sigma_0\, s^{-8}(1+z)(1-z)^{-4}\left[A(1+z)^{-2} + B\right]^2$$

where the values of A and B depend on the process and two parameters α and β

Reaction	A	B
$\pi^+p \to \pi^+p$	4α	β
$\pi^-p \to \pi^-p$	4β	α
$\pi^-p \to \pi^0n$	$2\sqrt{2}(\alpha+\beta)$	$(\alpha+\beta)/\sqrt{2}$
$K^+p \to K^+p$	4α	0
$K^-p \to K^-p$	0	α
$K_Lp \to K_Sp$	2β	$\beta/2$

All of these predictions are in good agreement with the data. The energy dependence is consistent with the data, which are not too accurate. The large angle behavior is also in quite good agreement with the valence quark model prediction of $\alpha=2\beta$ where α is the probability amplitude of finding a P-quark in the proton and β is the same for N-quark.

C. Annihilation and Transition Processes

If one claims to have a theory for the deep scattering region of a particular process then it is imperative to continue this result to the physical regions of the crossed reactions and see how well things turn out. Gluon exchange theories will have considerable difficulty in achieving even a qualitative agreement with experiment in this regard. The process $\bar{p}p \to \pi^- \pi^+$ can be gotten from the matrix element for $\pi^+ p \to \pi^+ p$ by the crossing $(s \leftrightarrow t)$. Using our previous matrix elements, one finds easily that

$$s^2 \frac{d\sigma}{dt} (\bar{p}p \to \pi^- \pi^+) = \frac{\sigma_0 (1-z^2)}{2s^6} \left[\alpha\ (1-z)^{-2} + \beta (1+z)^{-2}\right]^2$$

Notice that a forward and backward peak is predicted for the process and this is present in the experimental results of Chabaud *et al*, at 5 Gev/c. At 90°, we predict the ratio

$$\frac{d\sigma}{dt} (\bar{p}p \to \pi^- \pi^+) \Big/ \frac{d\sigma}{dt} (\pi^+ p \to \pi^+ p) = (\alpha+\beta)^2 \big/ 2(4\alpha+\beta)^2 \sim 1/18$$

Similar results hold for the reaction $(\bar{p}p \to K^- K^+)$ except that only a forward peak is predicted. This is again in agreement with experiment. The important case of $\bar{p}p \to \bar{p}p$ can be discussed using the form of the spin averaged matrix element given earlier. The result is that

$$\frac{d\sigma}{dt} (\bar{p}p \to \bar{p}p) \propto |u|^{0.4} \left(4|t|s\right)^{-5.2} \quad s^{-2}$$

which has only the forward peak expected from an (s,t) topology. Of some interest is the ratio

$$\frac{d\sigma}{dt} (pp) \Big/ \frac{d\sigma}{dt} (\bar{p}p) = \left(s/-u\right)^{5.6} \sim \left(2/1+z\right)^{5.6}$$

which is ∿50 at 90°. The experimental ratio at 5 Gev/c is approximately 100 ± 20.

It is also interesting that the interchange theory makes simple predictions for resonance production. Consider for example a process which changes a nucleon to a delta. If the delta wave function falls off in the same way as the nucleon, then the process will clearly have the same energy fall off as elastic scattering. If it falls off faster, then the momentum transfer will be routed through the nucleon vertex and the delta wave function will insure the convergence of the $k_\perp$ integral.

Again, the result is the same as the elastic case. Therefore, the transition form factors should scale with the nucleon elastic form factors and the process $p + p \to p + N^*$ should have the same dependence on energy and angle as elastic scattering. The data of Amaldi *et al* is in agreement with this prediction. Anderson *et al* have also shown that the reaction $\gamma + p \to \pi + \Delta$ has the same energy dependence as $\gamma + p \to \pi^+ + n$. The angular structure is also qualitatively the same at large angles.

V. Regge Effects

If the interchange theory is to make any pretense of being a complete dynamical theory, it must be able to be extended to smaller t values where Regge effects become important. If one assumes that the Regge region amplitude connects smoothly onto the deep region interchange amplitude, which is the most economical assumption to make and can be shown to be consistent with the most natural dynamics, then we must require that $M \sim (-u)^{\alpha(t)}\beta(t)$ for (-u) large and t fixed and $M \sim (-t)^{\alpha(u)}\beta(u)$ for (-t) large and u fixed. Using the approximate result

$$M \sim sF_A(-s)F_B(t)F_A(u)$$

one finds as $t \to \infty$:

$$\alpha(t) \sim 1-2A$$

$$\beta(t) \sim F_B(t) \sim (-t)^{-B}$$

and as $u \to \infty$:

$$\alpha(u) \sim 1-A-B$$

$$\beta(u) \sim F_A(u) \sim (-u)^{-A}$$

Therefore, not only are the limiting behaviors of the Regge function determined by the form factors but a priori unrelated Regge trajectories are not independent. In the case of pion-nucleon scattering, we see that the meson trajectory approaches the limit $\alpha(-\infty) \sim -1$ and the residue falls as $(-t)^{-2}$. The baryon trajectory which dominates for fixed u is seen to have the related limit $\alpha_n(-\infty) \sim -2$ (if spin is included one finds the natural result -3/2), and the residue falls as $(-u)^{-1}$. For pp scattering, one finds that the leading meson trajectory should have a different limit, $\alpha(-\infty) \sim -3$. This is different from the pion-nucleon case!

The way that the theory arranges the above behavior but retains factorization is quite simple. It is a mechanism which is found in the case of multichannel potential scattering but has not been discussed fully (if at all) before. In general, the dynamic forces more than one eigentrajectory to have the limiting behavior $\alpha(-\infty) \sim -1$. When these trajectories couple in the $\pi\pi$ and the πp case they add but in the pp channel they are equal and opposite in sign. They cancel and the leading behavior in pp scattering comes from a nonleading trajectory which has the limit $\alpha(-\infty) \sim -3$. This complicated behavior is forced upon the theory by the apparent difference in fall off of the pion and nucleon form factors. While it is a rather peculiar and unfamiliar behavior, it is still considerably better than the standard Regge picture of this large t region which is dominated by uncalculable branch cuts. The physical mechanism which leads to Regge behavior is hadronic bremsstrahlung as discussed in detail by Blankenbecler *et al.* If the basic interaction falls in energy, then it is preferable for one of the particles to emit a particle with a small fraction of the longitudinal momenta and then have it interact with the other particle (or its bremsstrahlung) at a lower effective energy. In an exclusive process, the emitted particles must be picked up and reabsorbed on the way out. Since this becomes more and more difficult as the momentum transfer increases, such Regge effects should disappear at large t and only the basic term survives. This behavior is actually found in a dynamical model which has been examined in detail. If the basic interaction is of the form

$$K(u,t) \sim (-u)^{-n}(-t)^{-m}$$

then the hadronic bremsstrahlung process leads to the Regge functions

$$\alpha(t) \cong -n+g\beta(t)/t + \ldots$$

and

$$\beta(t) \cong (-t)^{-m}$$

This type of behavior of $\alpha(t)$ and the limiting values of n and m discussed previously are consistent with a careful analysis of the available data by Coon *et al.* Further extension of the theory, especially to the coupled channel situation, will be found in Blankenbecler *et al*

VI. Inclusive Processes

Let us turn our attention to inclusive processes of the form A + B → C + X, where all possibilities for the particle or particles X are included in the cross section. Again, following Brodsky *et al*, our philosophy will be to find a kinematic regime where prediction can be made which are simply related to the short range behavior of the theory This again means that we do not attempt (at least not at first) to describe the most likely or typical event but only those rare events which are a direct reflection of the basic interactions at short distances. With that in mind, the natural diagram to examine, which is very similar to the one which worked well in exclusive scattering, is shown if Fig. 4. The variables are as before, s,t, and u with the addition of m^2, the $(\text{mass})^2$ of the system X. These kinematic variables

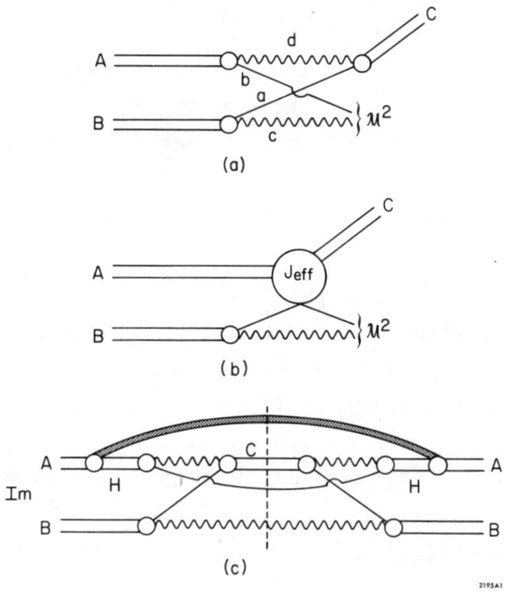

FIG. 4

will be defined in terms of the four momenta of the particles by $s = (A + B)^2$, $t = (A - C)^2$, $u = (B - C)^2$ and $m^2 = s+t+u - 3M^2$. The perpendicular momenta of particle C is given by $C_\perp{}^2 = tu/s$ and its Feynman (x_F) value is given by $x_F = (t-u)/s$.

It is convenient to work with the quantity R which is defined by

$$R = sC_0 \frac{d\sigma}{d^3C} = \frac{s^2}{\pi} \frac{d\sigma}{dt\,dm^2}$$

which is directly related to the scattering amplitude. For fixed x_F and $C_\perp$, Feynman has conjectured, and it seems to be well satisfied experimentally, that $R \sim sF(x_F, C_\perp)$ as s gets large. The contribution of the diagram shown in Fig (4a) is easily calculated, and one finds

$$R_0 = \frac{1}{2(2\pi)^3} \int \frac{d^2k_\perp dx}{x^3(1-x)^3} \psi_A^2\left(\vec{k}_\perp + (1-x)\vec{q}_\perp, x\right) \psi_B^2(\vec{k}_\perp, x) \psi_C^2(\vec{k}_\perp - x\vec{q}_\perp, x)$$

$$\times \Delta^2 \; \delta(E_i - E_f)$$

where

$$\Delta = E_{p+r} - E_a - E_d = -x\vec{r}_\perp^{\,2}(1-x)^{-1} + \ldots$$

$$E_i - E_f = s+u-(\vec{k}_\perp + \vec{q}_\perp)^2/x - k_\perp^2/(1-x)$$

$$\cong \frac{s+u}{x}\left(x + \frac{t}{s+u}\right)$$

and terms of order $(\text{mass})^2$ compared to s, t, u, and m^2 have been neglected. In the deep scattering region, R_0 becomes

$$R_0 \sim F_A^2(\frac{s}{s+u}\, p_\perp^2)\ F_C^2(\frac{s}{s+u}\, p_\perp^2)\ F_{2B}(x)$$

where $x = -t(s+u)^{-1}$ and we see that the structure function of the target, $F_{2B}(x)$ enters into R_0. It is quite reasonable that the inclusive cross section, in which hadron A goes into C and gives a momentum transfer t to break up the target, depends on $F_{2B}(x)$ which is the same process but is initiated by a photon.
As a example, one finds at fixed $C_\perp$ for the reaction $p+p \to p+X$, that

$$R_0\ (pp \to pX) \sim s^{-1}\ (\vec{C}_\perp^2)^{-6}$$

and hence Feynman scaling is not obeyed. This is not unexpected since the above result is expected to hold when $C_\perp^2$ is of the same order as s, and Feynman scaling holds at fixed $C_\perp$ as s goes to infinity. One should expect then that the inclusion of the effects of hadronic bremsstrahlung will build up R_0 and its energy dependence. In Fig (4c), the Reggeization process in which the incident particle A emits a slow hadron H with fractional longitudinal momentum z is illustrated.
This additional contribution to R, termed R_1, is easily computed to be of the form

$$R_1^{A+B\to C}(s,p_\perp^2,u) = \int_0^1 \frac{dz}{z} \sum_H G_{H/A}(z)\ R^{H+B\to C}(zs,p_\perp^2,u)\ ,$$

where a sum over all possible hadrons H is allowed. The z integral is limited by energy conservation and R_0 vanishes when the "effective" energy zs is not large enough to make particle C with transverse momenta $C_\perp$. As $C_\perp$ increases at fixed s, the lower limit on the z integral increases and the importance of R_1 and its Regge behavior relative to R_0 decreases.
The explicit calculation of these bremsstrahlung graphs shows that the Regge behavior shows up in the small z behavior of $G_{H/A}(z)$. If one writes this behavior as $G_{H/A}(z) \sim \beta_{H/A} z^{-\alpha_A(o)}$ then the explicit calculations show that $\alpha_A(o)$ is the value of the leading Regge trajectory at zero momentum transfer. The behavior of R_1 then reflects Regge behavior in the limit $x_\perp \equiv 2C_\perp/\sqrt{s} \to o$ but reflects the deep scattering behavior in the limit $x_\perp \sim$ finite. This very physical picture of the manner by which Regge behavior arises and disappears in various kinematic limits is one of the beauties of the interchange model. Finally, it sould be remarked that the target particle also undergoes hadronic bremsstrahlung and this must be included. If one uses the full expression for F_{2B}, which contains the effects of these processes, then they are included automatically.
In the limit of large s and $C_\perp$ the inclusive cross section can be written as

$$sC_0 \frac{d\sigma}{d^3C} = R_0(A+B\to C+X) + R_1(A+B\to C+X)+(A \leftrightarrow B)$$

where

$$R_0 = N_0 s^{\alpha_B} x_F^{1+\alpha_B}(1-x_F)^{\alpha_B-2(1-A-C)}(C_\perp^2)^{1-\alpha_B-2(A+C)}$$

and

$$R_1 = N_1 s^{\alpha_B} x_F^{1+\alpha_B} \int_{x_F}^{1} \frac{dz}{z} \sum G_{H/A}(z)\ (z-x_F)^{\alpha_B-2(1-A-C)}(zC_\perp^2)^{1-\alpha_B-2(H+C)}$$

The largest contribution to R_1 will be those hadrons H which have the smallest values of H, and this seems to be the mesons. Therefore, all hadronic interactions are denominated by meson-meson processes in the above kinematic regime.

It is interesting that the above form has the limiting behavior which arises in the Mueller analysis of inclusive reactions and provides many predictions on the form of the functions involved. The triple Regge region is defined by having t and m^2 very small compared s, and $x_F \sim 1$ where $(1-x_F)\sim m^2/s$. The above formula reduces to

$$R \sim s^{\alpha_B}(m^2/s)^{\alpha_B-2\alpha_{AC}(t)}\beta(t)$$

where $p_\perp^2 \sim -t$ and for $|t|$ larger than masses one predicts

$$\alpha_{AC}(t) \sim 1 - A - C$$

$$\beta(t) \sim (-t)^{2\alpha_{AC}(-\infty)-\alpha_B-1}.$$

These are very similar and agree with the earlier predictions about Regge behavior.

In the pionization region, where $t\sim u\sim -\sqrt{s}$ the Regge effects are very important and R_1 dominates R_0. A simple analysis shows that R achieves a simple form. If one simplifies the formula by setting $x_F = 0$, which implies that $t = u = -C_\perp\sqrt{s}$ then R becomes

$$R \sim R_1 = s^{\frac{1}{2}(\alpha_A+\alpha_B)} \sum_H \beta_{AH}\, \Gamma^H_{AC}(x_\perp)\ (C_\perp)^{2-\alpha_A-\alpha_B-4(H+C)}$$

where $\Gamma(x)$ is given by an integral over G and F_{2B}. If the trajectories are pomerons, then $\alpha_A=\alpha_B=1$ and Feynman scaling is achieved. Choosing H to be a meson then one predicts that

$$R \sim s\ (C_\perp)^{-4(1+C)}\Gamma_{AC}(x_\perp)$$

where the function $\Gamma(x_\perp)$ is smooth for small values of its arguments but vanishes as $x_\perp \sim 1$.

The prediction for the process $p + p \to \pi + X$ is $(C=1)$

$$C_0 \frac{d\sigma}{d^3C} \cong C_\perp^{-8}\ \Gamma(x_\perp) = s^{-4}\Gamma_1(x_\perp)$$

The scaling prediction varies as s^{-2} at fixed $x_\perp$ and if vector gluons are important they would lead to this prediction. In a very short time experimental results will be available which will choose between these theories. The latest results of Büsser *et al* seem to definitely prefer the interchange prediction. A variety of processes must be measured and compared with the predictions of the interchange theory before it can be fully accepted but it certainly looks as though it contains more than a modicum of truth at the present time

Let us briefly discuss some of the general features of the inclusive cross section as a function of $C_\perp$ at fixed x_F and energy. For large $C_\perp$ of order $s^{1/2}$, which is near the kinematical limit, the inclusive cross section should be dominated by diagrams of the type which gave rise to R_0. Therefore the cross section should not scale and should depend strongly on the particle types involved in the process. At small $C_\perp$ Feynman scaling should hold and the process will be s independent since the full effects of Reggeization via hadronic bremsstrahlung are kinematically allowed. Between these two limitng behaviors there is a transition region where the inclusive cross section should be given by a power fall off in $C_\perp$ times a function of $x_\perp$ which allows a smooth transition between the above two limits. As $x_\perp$ decreases, it becomes more and more complicated to compute the process in detail since more involved processes can occur at the upper vertex. For example, particle C may be produced along with other hadrons. To compute this would require quite a detailed knowledge of multi-hadron wave functions. However, it is just these processes which contribute to the Feynman scaling region of large multiplicity.

VII. Summary

The interchange theory provides a simple picture of large momentum transfer scattering which correlates many features of hadrons and their interactions. It is simple to compute and has considerable predictive power. It unites the electromagnetic structure of hadrons as expressed through their form factors and inelastic structure functions with elastic and inleastic hadron-hadron scattering. The thory joins smoothly onto Regge behavior which controls forward scattering and in fact predicts such behavior. The unified description of large and small momentum transfer scattering provided by the interchange model should allow considerable insight into the interaction of hadrons and their possible composite nature. It already yields a remarkably simple quantitative description which seems valid all the way from large angle elastic scattering at 5 Gev/c to inclusive scattering at the CERN-ISR.

References

V. Amaldi *et al*, Phys. Letters 34B, 435 (1971)

R. L. Anderson *et al*, Phys. Rev. Letters 30, 627 (1973)

V. Barger, F. Halzen and J. Luthe, Phys. Letters 42B, 428 (1972)

S. M. Berman, J. D. Bjorken and J. B. Kogut, Phys. Rev. D4, 3388 (1971)

R. Blankenbecler, S. J. Brodsky, J. F. Gunion and R. Savit, Phys. Rev. D8, 4117 (1973)

S. J. Brodsky, J. F. Gunion and R. Blankenbecler, Phys. Rev. D6, 2652 (1972), Phys. Letters 42B, 461 (1973)

V. Chabaud *et al*, Phys. Letters 41B, 209 (1972); F. W. Büsser *et al*, Phys. Letters 46B, 471 (1973)

F. E. Close, S. J. Brodsky, and J. F. Gunion, Phys. Rev. D8, 3678 (1973)

D. D. Coon, Tran Thanh Van, J. F. Gunion and R. Blankenbecler (unpublished)

J. F. Gunion, S. J. Brodsky and R. Blankenbecler, Phys. Rev. D8, 287 (1973), Phys. Letters 39B, 649 (1972)

P. V. Landshoff and J. C. Polkinghorne, Phys. Rev. D8, 927 (1973)

A. Muller, Phys. Rev. D2, 2693 (1970)

M. Schmidt, Phys. Rev. D9, 408 (1974)

KURT SYMANZIK
Deutsches Elektronen Synchrotron, Germany

SMALL-DISTANCE BEHAVIOUR IN FIELD THEORY

CONTENTS

KURT SYMANZIK
DESY, HAMBURG, FED. REP. GERMANY

SMALL-DISTANCE BEHAVIOUR IN FIELD THEORY

These lectures give a review of techniques for large-momenta-behaviour analysis in renormalizable field theories, and of results thereby obtained. For the sake of brevity of formulae, the ϕ^4 theory will be used to explain the ideas involved. Only in the last two sections more general theories will be discussed and specific results for some particular theories mentioned.

1. Renormalization.

As already mentioned, only the ϕ^4 theory will be discussed here, and for our purpose it suffices to do so on a heuristic level. For a more rigorous treatment see e.g. [1] and [2]. Some more details to the material of this section can be looked up in [3],[4], and, in particular, [5].

1.1 Generating functionals of Green's and vertex functions.

Let L be the Lagrangian density of ϕ^4 theory, given in detail in (1.11) below. The generating functional of the Green's functions containing connected and disconnected parts, namely, the vacuum expectation values of time-ordered operator products,

$$G_{disc}\{j\} = \langle (e^{i\int J(x)\phi(x)dx})_+ \rangle \tag{1.1a}$$

is obtained [6] as the scalar product

$$G_{disc}\{J\} = \overset{J}{\underset{out}{<}} \mid \overset{J}{\underset{in}{>}} \tag{1.1b}$$

of the in-and out-vacua of the theory with explicitly time dependent Lagrangian density

$$L^J = L + J\phi. \tag{1.1c}$$

The generating functional of the connected parts of the Green's functions $G_{disc}(x_1 \ldots x_n)$ is

$$G\{j\} = \ell n\, G_{disc}\{J\}$$

$$= \sum_{\substack{n=2 \\ n\ \text{even}}}^{\infty} \frac{i^n}{n!} \int \ldots \int dx_1 \ldots dx_n\ J(x_1) \ldots J(x_n) G(x_1 \ldots x_n) \tag{1.2}$$

where in the ϕ^4 model, in absence of spontaneous breaking of the $\phi\to-\phi$ symmetry, only the $G(x_1...x_n)$ and $G_{disc}(x_1.....x_n)$ with n even are nonzero. For functional derivatives we use the notation

$$\frac{\delta^n G\{J\}}{\delta J(x_1)...\delta J(x_n)} = G_{x_1...x_n}\{J\}$$

and similarly later for other functionals.

In renormalization theory more basic than the Green's functions are the amputated one-particle irreducible parts of the connected functions, $\Gamma(x_1...x_n)$, here called vertex functions (VFs). They are introduced most conveniently as follows [6], [7]: Set

$$\frac{1}{i} G_x\{J\} = \int dy\ i\Delta_F'(x-y)J(y) + O(J^3) \equiv A(x)\{J\}. \tag{1.3}$$

This can be solved to

$$J(x) = -i\int \Delta_F'^{-1}(x-y)A(y)dy + O(A^3). \tag{1.4}$$

The functional of A

$$\Gamma\{A\} = \left[G\{J\} - i\int dxA(x)J(x)\right]\Big|_{J=J\{A\}}. \tag{1.5}$$

satisfies

$$J(x) = i\,\frac{\delta\Gamma\{A\}}{\delta A(x)} \equiv i\Gamma_x\{A\} \tag{1.6}$$

and

$$\delta\Gamma\{A\}\Big|_{A\ \text{fixed}} = \delta G\{J\}\Big|_{J\ \text{fixed}}, \tag{1.7}$$

and the VFs are

$$\Gamma(x_1...x_n) = \Gamma_{x_1...x_n}\{0\}. \tag{1.8}$$

Their Fourier transforms are introduced by

$$\int dx_1...dx_n \Gamma(x_1...x_n)e^{i\Sigma p_i x_i} = (2\pi)^4\delta(\Sigma p_i)\Gamma(p_1...p_n)$$

and for n=2 we have [4]

$$\Gamma(p(-p)) = -G(p(-p))^{-1}. \tag{1.9}$$

1.2 Feynman rules, skeleton expansions, and renormalization conditions.

The Lagrangian L^J for ϕ^4 theory, written in unrenormalized fields ϕ_u (and with unrenormalized source J_u) is

$$L^J = \frac{1}{2}(\partial_\mu\phi_u\partial^\mu\phi_u - m_u^2\phi_u^2) - \frac{g_u}{4!}\phi_u^4 + J_u\phi_u . \tag{1.10}$$

Introducing renormalized fields, mass squared, coupling constant, and source function, by

$$\phi = Z_3^{-1/2}\,\phi_u \qquad m^2 = m_u^2 + \delta m^2$$

$$g = g_u Z_1^{-1} Z_3^2 \qquad J = Z_3^{1/2}\,J_u$$

the Lagrangian takes the form

$$L^J = \frac{1}{2}\,(\partial_\mu\phi\partial^\mu\phi - m^2\phi^2) - \frac{g}{4!}\,\phi^4 + J\phi$$

$$+(Z_3-1)\frac{1}{2}(\partial_\mu\phi\partial^\mu\phi - m^2\phi^2) + Z_3\delta m^2\,\frac{1}{2}\phi^2 - (Z_1-1)\,\frac{g}{4!}\,\phi^4 \tag{1.11}$$

It is discussed e.g. in [3] and [5] how from (1.11) the Feynman rules follow, the role of the counter terms as subtraction terms for the superficially divergent VFs emerges, and skeleton expansions are obtained. The Feynman rules are: Use $\Delta_F = -i(m^2-p^2-i\varepsilon)^{-1}$ for the propagators, $-ig$ for the vertices, and $(2\pi)^{-4}\int dk$ for each loop integration, and supply the symmetry factor S^{-1}, where S is the number of possible mappings of the graph on itself [2]. The superficial divergence degree D, i.e. divergence degree under overall power counting, is computed as follows: Let n_i be the number of ϕ legs and d_i the number of derivatives at the ith vertex, $\mathcal{L}$ the number of loops, L the one of lines, and E the one of external ϕ arguments amputated upon, equal n in (1.8). Then

$$D = 4\mathcal{L} - 2L + \sum_i d_i\ . \tag{1.12a}$$

Use of the familiar topological relation

$$\sum_i 1 + \mathcal{L} = L + 1 \tag{1.12b}$$

and of the trivial one

$$E = \sum_i n_i - 2L$$

in (1.12a) yields

$$D = 4 - E - \sum_i (4-n_i-d_i) \tag{1.13}$$

where $n_i + d_i$ is the ("operator") dimension of the ith vertex: the (mass) dimension of ϕ is 1, and the one of a derivative also 1. We thus have the table

E =	2	4	$\geq$ 6
D =	2	0	$\leq$ -2
graphical notation	$-\ \overline{-1}$		

The VFs for E = 2k, k $\geq$ 3, are expandible in terms of the renormalization functions: the propagator (the negative inverse VF for n=2) and the n=4 VF.

Example: E=6, D<0 :  = $\sum_{15}$ 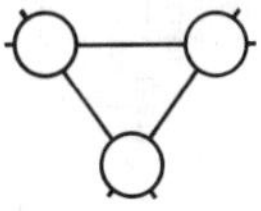 + $\sum_{30}$ 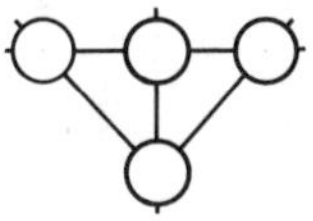 +...

For E=4 one finds, written diagrammatically

E=4, D=0  = X + $\frac{1}{6}$  + $\frac{1}{2}\sum_{3}$ 

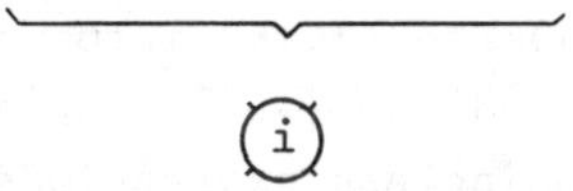

=  + $\frac{1}{2}$  (1.14)

where  is the Bethe-Salpeter (BS) kernel two-particle irreducible in the vertical channel, and itself expressed by

 =  + $\frac{1}{2}$ 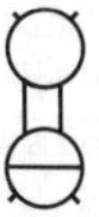 + $\frac{1}{2}$ 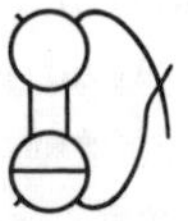 (1.15)

For E=2, the negative inverse propagator, one finds

E=2: D=2 $\quad - \overline{-1} = i(p^2-m^2) + \quad X \quad + \frac{1}{6}$ 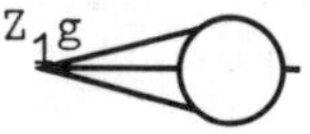 . (1.16)

Renormalization theory [8],[9] assures that to render integrals for superficailly divergent VFs finite, one must give as many data for these functions as a polynomial in momenta (with invariances observed) of order D has; the fulfilment of these renormalization conditions is effected by the counter terms indicated by crosses, which are $-iZ_1g$ in (1.14) and $i(Z_3-1)(p^2-m^2) + iZ_3\ \delta m^2$ in (1.16). In practice, it is simplest to subtract the Taylor expansion of order D at zero momenta from the sum of integrals for the function considered and to add that (invariant) polynomial of order D with finite coefficients that secures fulfilment of the renormalization conditions to the order in question. We choose

$$\Gamma(p(-p))\Big|_{p^2=m^2} = 0 \tag{1.17a}$$

$$\frac{\partial}{\partial p^2}\Gamma(p(-p))\Big|_{p^2=m^2} = i \tag{1.17b}$$

$$\Gamma(p_1 \ldots p_4)\Big|_{\underbrace{\text{symm.pt. to } m^2}_{p_i p_j = \frac{m^2}{3}(4\delta_{ij}-1)}} = -\,ig \tag{1.17c}$$

which imply

$$\Gamma(p(-p)) = i(p^2-m^2)+O(g^2) \tag{1.18a}$$

$$\Gamma(p_1 \ldots p_4) = -ig + O(g^2) \tag{1.18b}$$

$$\Gamma(p_1 \ldots p_{2n}) = O(g^n) \qquad n \geqq 3 \tag{1.18c}$$

For one procedure to convert (1.14) and (1.16) into manifestly finite form see [5], for a closely related one [3], for another [10]. Thereby, in intermediate steps meaningless expressions are manipulated; all calculations can be made meaningful, however, by regularizing the theory first e.g. as detailed in [3], and removing the regularization only in the final equations, in which all overlapping divergences are disentangled. The most elegant way to regularize, however, and of definite advantage in gauge field theories, is to perform all computations in $4-\varepsilon$ dimensions and to let $\varepsilon \to 0$ in the final expressions [11]: the Feynman rules lead upon integration to functions meromorphic in ε, and renormalization removes the poles at $\varepsilon = 0$.

1.3 Bethe-Salpeter equation.

For the BS equation that is part of (1.14) we use the matrix notation

$$I = B + IGB = B + BGI \tag{1.19a}$$

or equivalently

$$(1-GB)(1+GI)=(1+GI)(1-GB)=(1-BG)(1+IG)=(1+IG)(1-BG) = 1. \tag{1.19b}$$

Here I is the four-point VF, B the BS kernel, and G stands for the pair of propagators. Momenta are indicated as

$$\Gamma((-\tfrac{1}{2}q-p)\ (-\tfrac{1}{2}\,q+p)(\tfrac{1}{2}\,q+p')\ (\tfrac{1}{2}\,q-p')) \leftrightarrow p\overset{q}{I}p'$$

$$(2\pi)^4\left[\delta(p-p')+\delta(p+p')\right]G(-\tfrac{1}{2}\,q-p)G(-\tfrac{1}{2}\,q+p) \leftrightarrow \overset{q}{G}$$

$$(2\pi)^4[\delta(p-p') + \delta(P+p')] \leftrightarrow 1$$

and are suppressed if internal (and then integrated over with a factor $\frac{1}{2}(2\pi)^{-4}$ supplied), or if external (left relative, right relative, and through going or total) and kept general. In (1.19) it is understood that internal relative momenta fit unless a momentum associated with the factor to the left or right is indicated to be fixed, and thus constant in the internal-momentum integration, indicated for clarity additionally by a bracket > or < . For a G-link, this amounts to let arguments coalesce in coordinate space. B contains an ill-defined constant since the $-iZ_1g$ term in $\textcircled{i}$ in (1.14) is not appropriate to make B finite when the regularization is removed since it is chosen to render I finite instead. Thus, only B subtracted once in an arbitrary way is finite after regularization removal. We briefly indicate how I is obtained (e.g., recursively in perturbation theory) from the subtracted B. In the following, for simplicity of notation we write formulae as if the momenta set at which I is specified (as ${}^\circ\mathring{I}^\circ$) were all momenta vanishing. From (1.19a) follows

$${}^\circ\mathring{I} - {}^\circ\mathring{I}^\circ = {}^\circ\mathring{B} - {}^\circ\mathring{B}^\circ + {}^\circ\mathring{I}\,\mathring{G}\,(\mathring{B} - \mathring{B}^\circ) \tag{1.20}$$

which yields

$${}^\circ\mathring{I} = ({}^\circ\mathring{I}^\circ + {}^\circ\mathring{B} - {}^\circ\mathring{B}^\circ)\,[1 - \mathring{G}(\mathring{B} - \mathring{B}^\circ)]^{-1} . \tag{1.21}$$

The inverse defined e.g. by expansion (sufficient for perturbation theory). Similarly, as (1.21)

$$\mathring{I} = [1 - (\mathring{B} - {}^\circ\mathring{B})\,\mathring{G}]^{-1}\,({}^\circ\mathring{I} + \mathring{B} - {}^\circ\mathring{B}) \tag{1.22}$$

is derived, where ${}^\circ\mathring{I}$ from (1.21) is to be inserted. Finally, (2.2) below yields

$$I = [1 - (1+\mathring{I}\mathring{G})(B-\mathring{B})G - \mathring{I}(G-\mathring{G})]^{-1}\,[\mathring{I} + (1+\mathring{I}\mathring{G})(B-\mathring{B})] \tag{1.23}$$

which is, with $\mathring{I}$ from (1.22) inserted, the desired result. B itself can be recovered from

$${}^\circ\mathring{B}^\circ = \frac{{}^\circ\mathring{I}^\circ - {}^\circ\mathring{I}\mathring{G}(\mathring{B}^\circ - {}^\circ\mathring{B}^\circ)}{1 + {}^\circ\mathring{I}\mathring{G} >}$$

which is divergent in perturbation theory from order g^2 on when the regularization is removed. We shall also use the BS equation for the absorptive part of the four-point vertex. We introduce

$$p\,\overset{q}{I}\,p' + \overline{p\,\overset{q}{I}\,p'} \leftrightarrow p\,\overset{q}{\tilde{I}}\,p'$$

$$\text{for } (p-p')^2 \geq 4m^2, \quad p_0-p_0' \geq 2m, \quad (-\tfrac{1}{2}q-p)^2 < 9m^2 \ldots$$

$$\ldots (\tfrac{1}{2}q-p')^2 < 9m^2, \text{ and}$$

$$(2\pi)^4\delta(p-p')\,\overline{G}(-\tfrac{1}{2}q+p)G(-\tfrac{1}{2}q-p) \leftrightarrow \overset{q}{\tilde{G}},$$

$$(2\pi)^4\delta(p-p') \leftrightarrow \tilde{1},$$

and

$$p \overset{q}{B} p' + \overline{p \overset{q}{B} p'} + \ldots\ldots\ldots \leftrightarrow p \overset{q}{\tilde{V}} p'$$

where $\tilde{V}$ in the sum over all contributions to $\tilde{I}$ not $\tilde{G}$ - reducible in the horizontal channel. The equation

$$\tilde{I} = \tilde{V} + \tilde{I}\tilde{G}\tilde{V} = \tilde{V} + \tilde{V}\tilde{G}\tilde{I}, \tag{1.24a}$$

where in intermediate integrations the factor $\frac{1}{2}$ is to be dropped, was derived in [12] and can be recast into the forms

$$(\tilde{1} - \tilde{G}\tilde{V})(\tilde{1} + \tilde{G}\tilde{I}) = \ldots\ldots = 1 \tag{1.24b}$$

analogous to (1.19b). The ill-defined (imaginary) constant in B drops out in $\tilde{V}$.

2. Short-distance expansions.

Short distance (SD) expansions were proposed by Wilson and proven to all orders of perturbation theory by Zimmermann [2]. Here we give an elementary derivation of the simplest formulae that we will need later on. More details to this section can be found in the appendices A of [13] and [14].

2.1 Short-distance expansion in the narrow sense.

From (1.19a) follows

$$I - \mathring{I} = B - \mathring{B} + (I - \mathring{I})\,GB + \mathring{I}\,(G-\mathring{G})B + \mathring{I}\mathring{G}\,(B - \mathring{B}) \tag{2.1}$$

and therefrom with (1.19b)

$$I - \mathring{I} = (1+\mathring{I}\mathring{G})(B-\mathring{B})(1+GI) + \mathring{I}(G-\mathring{G})\,I. \tag{2.2}$$

Inserting this into the identity

$$GI = \mathring{G}\mathring{I} + (G - \mathring{G})I + \mathring{G}\,(I-\mathring{I})$$

yields

$$\begin{aligned} 1 + GI = {} & 1 + \mathring{G}\mathring{I} + (G - \mathring{G})\,I \\ & + (1 + \mathring{G}\mathring{I})\,\mathring{G}(B-\mathring{B})(1+GI) + \mathring{G}\mathring{I}(G-\mathring{G})I \end{aligned}$$

$$= (1+\mathring{G}\mathring{I}) \left[1+\mathring{G}(B-\mathring{B})(1+GI) + (G-\mathring{G})I \right] . \qquad (2.3)$$

The transposed form of (1.22)

$$\mathring{I} = (I^{\circ} + \mathring{B} - \mathring{B}^{\circ}) \left[1 - \mathring{G}(\mathring{B} - \mathring{B}^{\circ}) \right]^{-1} \qquad (2.4)$$

is equivalent to

$$1 + \mathring{G}\mathring{I} = (1 + \mathring{G}I^{\circ}) \left[1 - \mathring{G}(\mathring{B} - \mathring{B}^{\circ}) \right]^{-1} . \qquad (2.5)$$

Inserting (2.5) in (2.3) gives

$$1 + GI = (1 + \mathring{G}\, \mathring{I}^{\circ})\ (1 + W) \qquad (2.6)$$

where we used the abbreviation

$$W = -1 + \left[1-\mathring{G}(\mathring{B}-\mathring{B}^{\circ}) \right]^{-1} \left[1+\mathring{G}(B-\mathring{B})(1+GI) + (G-\mathring{G})I \right] . \qquad (2.7)$$

Acting with (2.6) on the suitably normalized bare vertex (a, upon regularization, finite constant in momentum space) gives

$$\gamma(1 + GI) = \Gamma = \gamma(1 + \mathring{G}\mathring{I}^{\circ})(1+W) = \mathring{\Gamma}^{\circ} < (1 + W) \qquad (2.8)$$

where we use the notation

$$\Gamma((\tfrac{1}{2}q+p)\ (\tfrac{1}{2}q-p),\ (-q)) \leftrightarrow \overset{q}{\Gamma}{}^{p} .$$

Hereby, $-q$ is the momentum acting on the (suitably normalized, see below) finite composite operator $1/2\ N_2(\phi^2)$ in the sense of Zimmermann [2]. In the present context, matrix elements of this operator are defined in terms of skeleton expansions in which the Γ just introduced appears, and for the normalization of which we choose Zimmermann's

$$\Gamma(00,0;\ m^2,g) = 1. \qquad (2.9)$$

Normalization independent, (2.6) takes the form

$$1 + GI = \mathring{G}\ \mathring{I}^{\circ} \frac{\Gamma}{\mathring{\Gamma}^{\circ}} + 1 + W . \qquad (2.10)$$

Using this in the skeleton expansion of a general matrix element, (2.10) is equivalent to

$$N_o(\phi(x+\xi)\phi(x-\xi)) = E(\xi)\ \tfrac{1}{2}\ N_2(\phi^2(x)) + N_2(\phi(x+\xi)(x-\xi)) \qquad (2.11)$$

in Zimmermann's notation [2]. In (2.10) and (2.11), the parts most singular as the left relative momentum goes to infinity, or $\xi \to 0$, respectively, is isolated and seen to be factorizable; the remainder term admits to be integrated over that relative momentum, or to let $\xi \to 0$, respectively, whereby the cofactor of the first term is again obtained. That that term, $<(1+W)$ in (2.8), is finite follows from examining all possibly divergent integrations involved in (2.7) and (2.8) and observing that they have a subtraction ("from outside") yielding convergence. The general matrix element of (2.11) corresponds to attaching to (2.10) from the right the diagrams

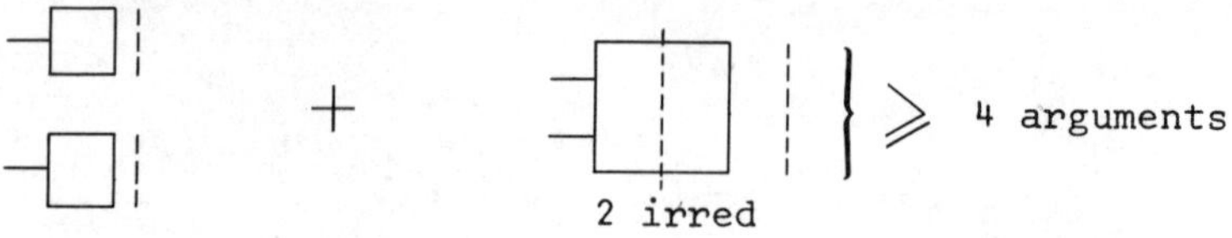

where the square functions are the general connected (amputated) functions with only the two-particle-irreducibility imposed. It is this condition that prevents a divergent integral to appear in the attaching process.

Using the function Γ constructed in (2.8) we can define Green's functions

$$<(\phi(x_1)\ldots\phi(x_{2n})\ \tfrac{1}{2}\ N_2(\phi^2(y_1))\ldots\ \tfrac{1}{2}\ N_2(\phi^2(y_\ell)))_+>_{conn.}$$

$$= G\ (x_1\ldots\ldots x_{2n},\ y_1\ldots y_\ell;\ m^2,g)$$

and corresponding vertex functions

$$\Gamma(x_1\ldots\ldots x_{2n},\ y_1\ldots\ldots y_\ell;\ m^2,g)$$

and Fourier transforms (with momentum-conserving delta function taken out, as before)

$$\Gamma(p_1\ldots p_{2n},\ q_1\ldots q_\ell;\ m^2,g)$$

by skeleton expansion, with superficial divergence, from (1.13),

$$D = \dim \Gamma = 4 - 2n - 2\ell. \tag{2.12}$$

In view of (2.12) and (2.9), only for $n = 0$, $\ell = 2$ must a further renormalization condition be imposed, and we choose

$$\Gamma(,\ 00;\ m^2,g) = 0\ . \tag{2.13}$$

The final, subtractive renormalization that ensures (2.13) is best imposed on the subtracted form

$$\Pi - \overset{\circ}{\Pi} = \Gamma(G-\overset{\circ}{G})\overset{\circ}{\Gamma}{}^T + \Gamma G(B-\overset{\circ}{B})\overset{\circ}{G}\overset{\circ}{\Gamma}{}^T \tag{2.14}$$

of the skeleton expansion for

$$\Gamma(,q(-q);\ m^2,g) \leftrightarrow \overset{q}{\Pi}$$

which is easily derived from (cp. (2.8))

$$\Pi = \gamma(G + GIG)\gamma^T + \text{const}$$
$$= \Gamma G\gamma^T + \text{const} = \gamma G\Gamma^T + \text{const}.$$

From (2.12) and (2.13) follow

$$\Gamma(p_1 p_2, q_1; m^2, g) = 1 + O(g) \quad , \tag{2.15a}$$

$$\Gamma(,q\ (-q); m^2, g) = O(1). \tag{2.15b}$$

2.2 Related expansion formulae.

(2.2) can be written

$$I=\mathring{I}\left[1+G(B-\mathring{B})(1+GI)+(G-\mathring{G})I\right] + (B-\mathring{B})(1+GI). \tag{2.16}$$

Inserting here (2.4) yields, with (2.7) and (2.8)

$$I=\mathring{I}^{\circ}(\mathring{\Gamma}^{\circ})^{-1}\Gamma +(\mathring{B}-\mathring{B}^{\circ})(1+W) + (B-\mathring{B})(1+GI) \tag{2.17}$$

which is, for large left relative momentum, a SD expansion for an amputated function in contrast to (2.10).

From (2.8) and (1.19) follows the subtracted BS equation for Γ

$$\Gamma-\mathring{\Gamma} = \Gamma GB - \mathring{\Gamma}\mathring{G}\mathring{B} = (\Gamma-\mathring{\Gamma})\mathring{G}\mathring{B}+\Gamma(G-\mathring{G})\mathring{B} + \Gamma G(B-\mathring{B}\ )$$
$$= \Gamma(G-\mathring{G})\mathring{I} + \Gamma G(B-\mathring{B})(1+\mathring{G}\mathring{I})$$

and herefrom, using (2.17) at zero total momentum in transposed form,

$$\Gamma=\mathring{\Gamma} + (\Pi-\mathring{\Pi})(\mathring{\Gamma}^{\circ})^{-1}\ {}^{\circ}\mathring{I} + \Gamma R \tag{2.18}$$

where we have used (2.14) and the abbreviation

$$R=G(B-\mathring{B}) + \left[G-\mathring{G} + G(B-\mathring{B})\mathring{G}\right](1+\mathring{W}^{T})(\mathring{B}-{}^{\circ}\mathring{B}).$$

(2.18) is an expansion formula for large (right) relative momentum. Inserting (2.18) into (2.17) subtracted from itself at zero momentum yields

$$I = \mathring{I} + \mathring{I}^{\circ}(\mathring{\Gamma}^{\circ})^{-1}\ (\Pi-\mathring{\Pi})(\mathring{\Gamma}^{\circ})^{-1}\ {}^{\circ}\mathring{I} \tag{2.19}$$
$$+ (\mathring{B}-\mathring{B}^{\circ})(W-\mathring{W}) + (B-\mathring{B})(1+GI) + \mathring{I}^{\circ}(\mathring{\Gamma}^{\circ})^{-1}\Gamma R\ ,$$

an expansion formula for both the left and right relative momentum large compared to the total momentum. (2.17-19) will be used in sect. 5.

We finally derive an expansion formula for absorptive parts used recently by Mueller [15]. From (1.24) follows, by the same algebraic steps, the analog of (2.17), for zero total momentum

$$\mathring{\tilde{I}} = \mathring{\tilde{I}}^{\circ}\ (\mathring{\Gamma}^{\circ})^{-1}\ \mathring{\tilde{\Gamma}} + (\mathring{\tilde{V}} - \mathring{\tilde{V}}^{\circ})(1 + \mathring{\tilde{W}}) \tag{2.20a}$$
$$\equiv \mathring{\tilde{I}}^{\circ}\ \mathring{\tilde{\Gamma}}_M + \mathring{\tilde{I}}^{\,reg}$$

with

$$\overset{\circ}{\tilde{\Gamma}}_M = <(1 + \overset{\circ}{\tilde{W}}) \tag{2.20b}$$

and

$$1 + \overset{\circ}{\tilde{W}} = [1 - \overset{\circ}{\tilde{G}} (\overset{\circ}{\tilde{V}} - \tilde{V}^\circ)]^{-1}. \tag{2.20c}$$

$\tilde{I}^{reg}$ vanishes, like the analogous term in (2.17), for large left relative momentum squared. Actually, Mueller [15] first separates (1.24) for zero total momentum with respect to a conserved angular quantum number, whereupon (1.24a) becomes a one-dimensional integral equation, with integration over the square of the intermediate relative momentum. (2.20) is then to be read for fixed angular quantum number with, where indicated, the right relative momentum squared zero rather than zero as a vector. The vanishing of $\overset{\circ}{\tilde{I}}^{reg}$ hereby for large left relative momentum squared is nontrivial.

3. Mass vertex insertion.

The references for this section are [16], [13], and, in particular, [14].

3.1 Derivation of PDEs.

Consider in the computation of all Green's and vertex functions from the Lagrangian (1.11) (amended also by the source term $\frac{1}{2} KN_2(\phi^2)$ to generate $\frac{1}{2} N_2(\phi^2)$ - arguments) the replacement

$$L \to L - s \frac{1}{2} m^2 N_2(\phi^2)\phi(g) \equiv L^s \tag{3.1}$$

where

$$\phi(g) = 1 + O(g^2) \tag{3.2}$$

will be specified later. The generating functional of the Green's functions

$$G_{disc} \{J,K\} = <(e^{i\int J\phi + \frac{i}{2}\int KN_2(\phi^2)})_+ >$$

then changes accordingly:

$$G^s_{disc} \{J,K\}(m^2,g) = \text{const. } G_{disc} \{J,K-sm^2\phi(g)\} (m^2,g)$$

$$= G_{disc}\{(1+\Delta z(s))J,(1+\Delta u(s))K\}(m^2+\Delta m^2(s),g+\Delta g(s))$$

$$\cdot \exp [- \frac{i}{2} \Delta v(s) \int K^2(x)\, dx] \tag{3.3}$$

where the first equality is a consequence of the integrated form of the Schwinger action principle [6], and the second equality expresses that the change (3.1) is one within the familiy of ϕ^4 theories which are completely parametrized by the renormalization parameters m^2 and g, that (1.17b) and (2.12), to be upheld, require further multiplicative factors to be allowed, and that (2.13), to be upheld, may require to change the constant in (2.14) and thus to provide for an extra subtraction term for the n=0, ℓ=2 VF.

Since $N_2(\phi^2)$ is a finite operator, all $\Delta z(S)$, $\Delta u(S)$ etc. in (3.3) are finite and, in fact, computable in renormalized perturbation theory, see appendix C of [13] where, however, the formula following (C.7) should read

$$s = m^{-2}\,\phi(g)^{-1} \int_{m^2}^{m^2(s)} dm^2(s')\, Z(s')^{-1}\, \phi(g(s'))$$

$$\cdot \exp\Big[- \int_{m^2}^{m^2(s')} m^2(s'')^{-1}\, dm^2(s'')\,\eta(g(s''))\Big] .$$

Differentiating (3.3) with respect to s at s=0 gives with (1.2)

$$-im^2\phi(g)\,\frac{1}{i}\int dx\, \frac{\delta}{\delta K(x)}\, G\{J,K\}\,(m^2,g)$$

$$= \Big[\gamma(g)\int dx\, J(x)\,\frac{\delta}{\delta J(x)} + (2\gamma(g)+\eta(g))\int dx K(x)\,\frac{\delta}{\delta K(x)}$$

$$+ m^2\,\frac{\partial}{\partial m^2} + \beta(g)\,\frac{\partial}{\partial g}\Big] G\{J,K\}\,(m^2,g) - i\kappa(g)\,\frac{1}{2}\int K(x)^2\, dx \qquad (3.4)$$

where we have introduced the functions

$$\frac{\partial \Delta m^2(s)}{\partial s}\Big|_{s=0} = m^2 \quad \text{by normalization } (\phi(g) \text{ choice}) \qquad (3.5a)$$

$$\frac{\partial \Delta g^{(s)}}{\partial s}\Big|_{s=0} = \beta(g) \qquad (3.5b)$$

$$\frac{\partial \Delta z(s)}{\partial s}\Big|_{s=0} = \gamma(g) \qquad (3.5c)$$

$$\frac{\partial \Delta u(s)}{\partial s}\Big|_{s=0} = 2\gamma(g) + \eta(g) \qquad (3.5d)$$

$$\frac{\partial \Delta v(s)}{\partial s}\Big|_{s=0} = \kappa(g). \qquad (3.5e)$$

Using now (1.3), (1.6) and (1.7) gives

$$-im^2\phi(g)\,\frac{1}{i}\int dx\, \frac{\delta}{\delta K(x)}\, \Gamma\{A,\, K\}\,(m^2,g)$$

$$= \Big[-\gamma(g)\int dx\, A(x)\,\frac{\delta}{\delta A(x)} + (2\gamma(g)+\eta(g))\int dx\, K(x)\,\frac{\delta}{\delta K(x)}$$

$$+ m^2\,\frac{\partial}{\partial m^2} + \beta(g)\,\frac{\partial}{\partial g}\Big]\,\Gamma\,\{A,K\}\,(m^2,g) -$$

$$-\frac{i}{2}\,\kappa(g)\int dx\; K(x)^2. \tag{3.6}$$

Or, going from the generating functional to the VFs themselves,

$$O_{p_{2n},\ell}\Gamma(p_1\ldots p_{2n},q_1\ldots q_\ell;m^2,g) = -im^2\phi(g)\Gamma(p_1\ldots p_{2n},q_1\ldots q_\ell 0;m^2,g) -i\kappa(g)\delta_{no}\delta_{\ell 2} \tag{3.7a}$$

$$O\,p_{2n,\ell} \equiv m^2\frac{\partial}{\partial m^2} + \beta(g)\frac{\partial}{\partial g} - 2n\gamma(g) + \ell(2\gamma(g)+\eta(g)). \tag{3.7b}$$

Requiring consistency of (3.7) with the five renormalization conditions (1.17a-c), (2.9), (2.13) allows to compute the five functions in (3.5), e.g. as power series in g:

$$\beta(g) = b_0 g^2 + b_1 g^3 + \ldots \qquad b_0 = 3(32\pi^2)^{-1} > 0,\quad b_1 = -17(2^9 3\pi^4)^{-1} \tag{3.8a}$$

$$\gamma(g) = c_0 g^2 + c_1 g^3 + \ldots \qquad c_0 = (2^{11} 3\pi^4)^{-1} \tag{3.8b}$$

$$\eta(g) = h_0 g + h_1 g^2 + \ldots \qquad h_0 = \frac{1}{3} b_0 \tag{3.8c}$$

$$\kappa(g) = \kappa_0 + \kappa_1 g + \ldots \qquad \kappa_0 = \frac{1}{3} b_0 \tag{3.8d}$$

$$\phi(g) = 1 + f_2 g^2 + \ldots \tag{3.8e}$$

The value of b_1 was computed in [17].
Consider replacing the renormalization conditions (1.17b-c), (2.9), (2.13), by others such that (1.18a-b) and (2.15a-b) remain valid for the new functions $\overline{\Gamma}$ with $\overline{g}$ replacing g, e.g. replacing (1.17b-c) by

$$\overline{\Gamma}(00,\ ;m^2,\overline{g}) = -im^2$$

$$\overline{\Gamma}(0000\ ,\ ;\ m^2,\overline{g}) = -i\overline{g} \text{ etc.} \tag{3.9}$$

Then renormalization theory [9] assures that

$$\overline{\Gamma}(p_1\ldots p_{2n},\ q_1\ldots q_\ell;m^2,\overline{g}) = \overline{Z}_1(\overline{g})^n\,\overline{Z}_2(\overline{g})^\ell\,\Gamma(p_1\ldots p_{2n},q_1\ldots q_\ell;m^2,g\,(\overline{g})) + i\,\delta_{no}\delta_{\ell 2}\,\overline{f}(\overline{g}), \tag{3.10}$$

must hold, with, according to (1.18a-b) and (2.15a-b), power series

$$\overline{Z}_1(\overline{g}) = 1 + 0(\overline{g}^2) \tag{3.11a}$$

$$\overline{Z}_2(\overline{g}) = 1 + 0(\overline{g}) \tag{3.11b}$$

$$g(\overline{g}) = \overline{g} + 0(\overline{g}^2) \tag{3.11c}$$

$$\overline{f}(\overline{g}) = 0(1). \tag{3.11d}$$

Inserting (3.10) into the PDEs for $\overline{\Gamma}$, analogous to (3.7), gives relations between the new functions $\overline{\beta}(\overline{g})$ etc. and the ones in (3.7), e.g.

$$\beta(g(\overline{g}))^{-1}\, dg = \overline{\beta}(\overline{g})^{-1}\, d\overline{g}. \qquad (3.12)$$

This relation is consistent with (3.11c) and the analog of (3.8a),

$$\overline{\beta}(\overline{g}) = \overline{b}_0 \overline{g}^2 + \overline{b}_1\, \overline{g}^3 + \ldots ,$$

only if

$$\overline{b}_0 = b_0,\ \overline{b}_1 = b_1 . \qquad (3.13a)$$

Similarly, one derives

$$\overline{c}_0 = c_0,\ \overline{h}_0 = h_0\ ,\ \overline{\kappa}_0 = \kappa_0 . \qquad (3.13b)$$

Use of conventions (3.9) and of (3.13a) simplifies somewhat the computation of b_0 and b_1. If only $\beta(g)$ and $\overline{\beta}(\overline{g})$ are given, $g(\overline{g})$ in (3.11c) is obtainable from (3.12) only if in addition an integration constant, easily obtained from (3.10), is known.

3.2 Integration of PDEs.

The ordinary DE of the characteristic of the PDEs (3.7) is

$$\frac{dm^2}{m^2} = \frac{dg}{\beta(g)} .$$

The solution we parametrize by λ:

$$m^2, g \rightarrow m^2\lambda^2,\ g(\lambda)$$

where $g(\lambda)$ is defined by

$$\ell n\, \lambda^2 = \int_g^{g(\lambda)} dg'\ \beta(g')^{-1}. \qquad (3.14)$$

or, equivalently, by

$$g(\lambda) = \rho^{-1}\, (\ell n\lambda^2 + \rho(g)\ ) \qquad (3.15a)$$

where

$$\int^g \frac{dg'}{\beta(g')} = \rho(g). \qquad (3.15b)$$

From (3.15) follows

$$F(g(\lambda)) = F(\rho^{-1}(\ell n\lambda^2+\rho(s))) = \sum_{r=0}^{\infty} \frac{(\ell n\lambda^2)^r}{r!} \left[\frac{\partial}{\partial\rho(g)}\right]^r F(\rho^{-1}(\rho(g)))$$

$$= \sum_{r=0}^{\infty} \frac{(\ell n\lambda^2)^r}{r!} \left[\beta(g)\,\frac{\partial}{\partial g}\right]^r F(g) \qquad (3.16)$$

which yields the perturbation expansion of e.g. $g(\lambda)$ upon setting $F(g) \equiv g$ and using (3.8a). From (3.14) follows that $g(\lambda)$ increases or decreases as a function of increasing λ if $\beta(g(\lambda))$ is positive or negative, respectively. In view of (3.8a) we obtain Fig. 1 where it is assumed that $\beta(g)$ has for positive g a first (and first-order) zero

Fig. 1

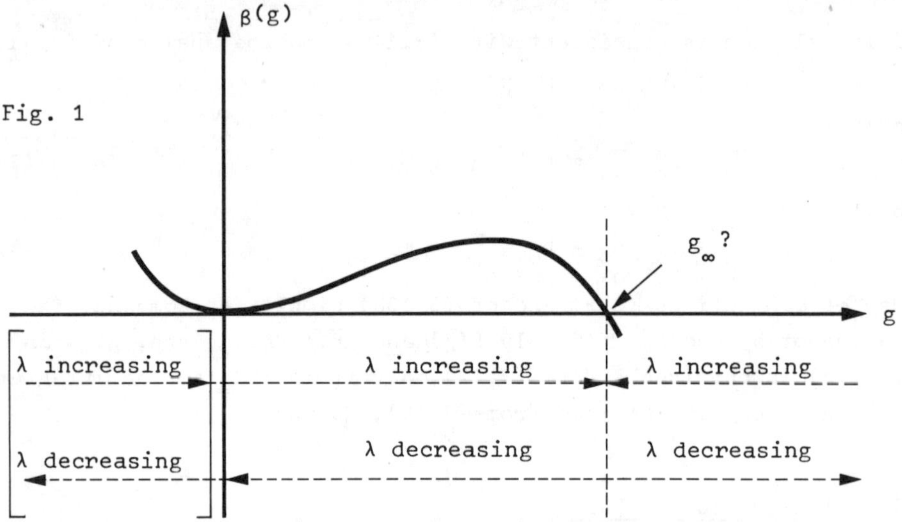

at g_∞. The existence of such a zero is uncertain, but the fact [17] that $b_1 < 0$ suggests it. For $g < 0$, the ϕ^4 theory can be argued to be unstable [18]. For $0 < g < g_\infty$, (3.15) gives

$$g(\lambda) = \frac{1}{b_o \ln(\lambda^{-2})} - \frac{b_1 \ln\ln\lambda^{-2}}{b_o^3(\ln \lambda^{-2})^2} + 0\left(\frac{1}{(\ln\lambda)^2}\right), \quad \lambda\to 0. \tag{3.17}$$

We now introduce the functions

$$a(g) = \exp\left[2\int_o^g \frac{dg'}{\beta(g')}\gamma(g')\right] = 1+2\frac{c_o}{b_o}g + \ldots \tag{3.18a}$$

$$h(g) = \exp\left[\int_o^g \frac{dg'}{\beta(g')}\eta(g')\right] = g^{1/3}\left[1 + \ldots g + \ldots\right] \tag{3.18b}$$

$$k(g)= \int^g \frac{dg'}{\beta(g')} a(g')^2 h(g')^2 \kappa(g') = -g^{-1/3}\left[1+\ldots g+\ldots\right] \tag{3.18c}$$

where in (3.18b) and (3.18c) the integration constants have been chosen conveniently. The integral of (3.7) is now

$$\begin{aligned}\Gamma(p_1\ldots p_{2n}, q_1\ldots q_\ell; m^2, g) &\\ = a(g)^{n-\ell}\, a(g(\lambda))^{-n+\ell}\, h(g)^{-\ell}\, h(g(\lambda))^{\ell} &\\ \cdot\, \Gamma(p_1\ldots p_{2n}, q_1\ldots q_\ell; m^2\lambda^2, g(\lambda)) -\end{aligned}$$

$$-i\delta_{no}\,\delta_{\ell 2}\,a(g)^{-2}h(g)^{-2}\left[k(g) - k(g(\lambda))\right]$$

$$-im^2 \int_{\lambda^2}^{1} d\lambda'^2\, a(g)^{n-\ell}\, a(g(\lambda'))^{-n+\ell}\, h(g)^{-\ell} h(g(\lambda'))^{\ell} \phi(g(\lambda'))$$

$$\cdot\, \Gamma(p_1 \ldots p_{2n},\, q_1 \ldots q_\ell\; 0;\; m^2\lambda'^2,\; g(\lambda')) \qquad (3.19)$$

where the sum of the first two terms on the r.h.s. can also be written as $\Gamma^{s(g,\lambda)}(p_1 \ldots p_{2n},\, q_1 \ldots q_\ell;\, m^2, g)$, the superscript meant in the sense of (3.1) and (3.3), (see Appendix C of [13], with, however, the correction of formulae as indicated above) for

$$s(g,\lambda) = \phi(g)^{-1} a(g) h(g) \int_1^{\lambda^2} d\lambda'^2 \phi(g(\lambda')) a(g(\lambda'))^{-1} h(g(\lambda'))^{-1}. \qquad (3.20)$$

Thus, (3.19) expresses the effect of noninfinitesimal mass vertex insertion as the integral over the infinitesimal steps thereto. The point of (3.19) is that the effects of the infinitesimal steps are more easily controlled than the effect of the noninfinitesimal operation.

Generalizing (3.1) one may also consider the effect of insertions of other finite operators, $N_4(\phi^4)$ and $N_4(\partial_\mu \phi \partial^\mu \phi)$, that occur in Zimmermann's [2] effective Lagrangian, and is led to generalizations [19] of the PDEs (3.7). However, these other PDEs have found no application yet since the vertex insertions then have operator dimension four, and VFs with such insertions do not have a characteristic property distinguishing them from the original VFs.

4. Asymptotic forms.

The references for this section are [16], [20], [13], and [14].

4.1 Existence of asymptotic forms.

(3.19) can be made useful only if λ can be chosen such that the r.h.s., excepting the integral, simplifies. In view of (3.17), the appropriate choice is $\lambda=0$.

From now on, we will consider, unless said otherwise, Euclidean momenta only, i.e. energy components (p_{i0}, q_{j0}) imaginary (ip_{i4}, iq_{j4}), space components real. Now from (2.12)

$$\Gamma(p_1 \ldots p_{2n}, q_1 \ldots q_\ell 0;\; m^2\lambda^2,\; \bar{g})$$

$$= \lambda^{-2+2n+2\ell}\,\Gamma(\lambda^{-1}p_1 \ldots \lambda^{-1}p_{2n},\; \lambda^{-1} q_1 \ldots \lambda^{-1} q_\ell 0; m^2, \bar{g}) \qquad (4.1)$$

and Weinberg's theorem [21] yields for the VF on the r.h.s. as $\lambda\to 0$ a behaviour proportional to $(\lambda^{-1})^{4-2n-\lambda(\ell+1)}(\ell n\lambda)^c$, with c depending on the order considered, provided the external momenta are not such that there exists a partition of the graph into subgraphs such that the external large momenta can be contained in some subgraphs and the sum of Weinberg exponents [21] for these subgraphs is larger than the exponent for the whole graph.

In the present model, it is required that there exist a partition of the external momenta into at least three nontrivial even sets (i.e. an even number of p and an arbitrary number of q momenta) of zero sum each, to obtain a Weinberg exponent larger than the superficial one. In view of the vanishing q-momentum in (3.21), the vanishing of one even partial sum of other momenta is required to invalidate the estimate mentioned before. Such momenta sets we call exceptional, other momenta sets (e.g., generic ones) nonexceptional. In formulae, nonexceptional momenta are defined by the condition that

$$\Gamma(p_1\dots p_{2n},\ q_1\dots q_\ell 0;m^2\lambda^2,\bar{g}) = O((\ell n\lambda)^{\cdots}) \tag{4.2}$$

at least to all orders in perturbation theory, and the necessary and sufficient condition (for Euclidean momenta) for (4.2) to hold was stated before.

If (4.2) holds, then due to (3.16) the integral in (3.19) will allow $\lambda\to 0$, and thus

$$\begin{aligned}&\lim_{\lambda\to 0}\Big[a(g)^{n-\ell}a(g(\lambda))^{-n+\ell}\ h(g)^{-\ell}h(g(\lambda))^{\ell}\ \Gamma(p_1..p_{2n},q_1..q_\ell;m^2\lambda^2,g(\lambda))\\ &-ia\,(g)^{-2}\ h(g)^{-2}\ \big[k(g)-k(g(\lambda))\big]\ \delta_{no}\delta_{\ell 2}\ \Big]\\ &\equiv \Gamma_{as}(p_1\dots p_{2n},\ q_1\dots q_\ell;m^2,g)\end{aligned} \tag{4.3}$$

then exists at least to all orders in perturbation theory; (3.17) suggests that (in this nonexceptional case) (4.3) exists also outside of perturbation theory. In view of what was said after (3.19), we have also

$$\begin{aligned}&\Gamma_{as}(p_1\dots p_{2n},\ q_1\dots q_\ell;m^2,g)\\ &\quad=\Gamma^{s(g,0)}(p_1\dots p_{2n},\ q_1\dots q_\ell;m^2,g)\end{aligned} \tag{4.4}$$

which identifies these functions as VFs of a ϕ^4 theory with that particular choice of the bare mass which, as we shall prove below, leads to a vanishing renormalized mass. We call the VFs Γ_{as} the preasymptotic ones.

4.2 Properties of asymptotic forms.

Replacing in (4.3) λ by $\lambda\lambda'$ and letting $\lambda'\to 0$ gives, due to $g(\lambda\lambda') = [g(\lambda)](\lambda')$ from (3.14),

$$a(g)^{n-\ell}a(g(\lambda))^{-n+\ell}h(g)^{-\ell}h(g(\lambda))^{\ell}\Gamma_{as}(p_1..p_{2n},q_1..q_\ell;m^2\lambda^2,g(\lambda))$$

$$-ia(g)^{-2}\, h(g)^{-2}[k(g) - k(g(\lambda))]\delta_{no}\delta_{\ell 2}$$

$$= \Gamma_{as}\,(p_1...p_{2n},\, q_1...q_\ell;m^2,\, g) \qquad (4.5)$$

which is equivalent to

$$Op_{2n,\ell}\Gamma_{as}(p_1...p_{2n},q_1...q_\ell;\, m^2,g)$$

$$=-\kappa(g)\;\delta_{no}\delta_{\ell 2} \qquad (4.6)$$

with $Op_{2n,\ell}$ from (3.7b). Rewriting (4.5) as

$$\Gamma_{as}(\lambda p_1...\lambda p_{2n},\lambda q_1...\lambda q_\ell;m^2,g)$$

$$= \lambda^{4-2n-2\ell}a(g)^{n-\ell}a(g(\lambda))^{-n+\ell}\; h(g)^{-\ell}h(g(\lambda))^{\ell}$$

$$\cdot\Gamma_{as}(p_1...p_{2n},\, q_1...q_\ell;m^2,g(\lambda))$$

$$-ia(g)^{-2}\, h(g)^{-2}[k(g) - k(g(\lambda))]\delta_{no}\delta_{\ell 2} \qquad (4.7)$$

allows, because of (3.17), to read off the small-momenta behaviour of Γ_{as}, to arbitrary precision apart from the λ-independent overall factor, from the series expansions of the functions of $g(\lambda)$ involved. In particular, it follows that for $p^2\to 0$

$$\Gamma_{as}(p(-p),\; ;m^2,g)$$

$$=ia(g)p^2[1-2b_o^{-2}\; c_o(\ell n[m^2(-p^2-i\varepsilon)^{-1}])^{-1}+0((\ell np^2)^{-2}\ell n\ell np^2], \qquad (4.8)$$

showing that, as claimed, the Γ_{as} describe a zero-mass theory, with moreover the massless particles discrete, a non-perturbation theoretical result. One may also say that, in view of (4.7) and $g(\lambda)\to 0$ as $\lambda\to 0$, the preasymptotic theory is for $0<g<g_\infty$ "asymptotically free in the infrared".

From (3.19) for $\lambda\to 0$, (4.3), and (2.12) follows

$$\Gamma(\lambda p_1...\lambda p_{2n},\lambda q_1...\lambda q_\ell;\; m^2,g)$$

$$=\Gamma_{as}\;(\lambda p_1...\lambda p_{2n},\lambda q_1...\lambda q_\ell;\; m^2,g)$$

$$-\; im^2\lambda^{2-2n-2\ell}\int_0^1 d\lambda'^2 a(g)^{n-\ell}a(g(\lambda'))^{-n+\ell}$$

$$\cdot\, h(g)^{-\ell}\, h(g(\lambda'))^{\ell}\phi(g(\lambda'))$$

$$\cdot\, \Gamma(p_1\ldots p_{2n},\, q_1\ldots q_\ell\; 0;\; m^2\lambda^{-2}\lambda'^2,\; g(\lambda')). \tag{4.9}$$

This shows that the Γ_{as} on the r.h.s. collects, in view of (4.2), and (3.16), for $\lambda\to\infty$ in perturbation theory all terms with factors $\lambda^{4-2n-2\ell}(\ell n\lambda)^K$, $K\geq 0$, and (4.7) shows that it collects these terms only while the remainder integral in (4.9) collects all terms with factors $\lambda^{2r-2n-2\ell}(\ell n\lambda^2)^K$, $K\geq 0$, $1\geq r > -\infty$.
Use of (3.19) in (4.3) allows to rewrite the limitand as

$$\Gamma_{as}(p_1\ldots p_{2n},\, q_1\ldots q_\ell;\; m^2,\; g)$$

$$-im^2\int_0^\lambda d\lambda'^2\; a(g)^{n-\ell}a(g(\lambda'))^{-n+\ell}h(g)^{-\ell}h(g(\lambda'))^{\ell}$$

$$\cdot\phi(g(\lambda'))\Gamma(p_1\ldots p_{2n},\, q_1\ldots q_\ell\; 0;\; m^2\lambda'^2,\; g(\lambda'))$$

where, due to (3.16) and (4.2), the integral yields in its perturbation theoretical expansion only terms $\lambda^{2r}\;(\ell n\lambda^2)^K$, $K\geq 0$, $r\geq 1$. It follows that in the limitand in (4.3) all terms not of the last-mentioned form, except λ-independent ones which yield the Γ_{as} function, cancel identically. This implies that in the limitand in (4.3), only the λ-independent terms need be treated precisely, all terms involving positive powers of $\ell n\lambda^2$ being immaterial, which in turn, in conjuction with our conclusions from (4.9) proves the correctness of the "elementary recipe": If $\Gamma(p_1\ldots p_{2n},q_1\ldots q_\ell;m^2\lambda^2,g)$ is expanded for $\lambda\to 0$ in a double power series in λ^2 and $\ell n\lambda^2$, then the sum of the λ-independent terms is $\Gamma_{as}\;(p_1\ldots p_{2n},\, q_1\ldots q_\ell;\; m^2,g)$, and the sum of the terms with factors $\lambda^{\circ}(\ell n\lambda^2)^K$, $K\geq 0$, is $\Gamma_{as}(p_1\ldots p_{2n},\, q_1\ldots q_\ell;\; m^2\lambda^2,\; g)$. To obtain from (4.9) all terms with factors of form $\lambda^{2-2n-2\ell}(\ell n\lambda^2)^K$, $K\geq 0$, requires the large-λ analysis of the VF in the integrand. These momenta are exceptional, however, such that (4.9) itself is not applicable for that VF. The appropriate modification of (4.9) will be discussed in Sect. 5.3.

4.3 Relation to zero-mass theory.

The zero-mass ϕ^4 theory VFs are directly constructed in perturbation theory in appropriate parametrization, analogous to the procedure of

Gell-Mann and Low [22] in QED, employing a normalization mass U and coupling constant, V, and denoted by $\Gamma_o(p_1 \dots p_{2n}, q_1 \dots q_\ell ; U^2, V)$. As renormalization conditions we choose

$$\Gamma_o(00, \; ; U^2, V) = 0, \tag{4.10a}$$

$$\Gamma_o(p(-p), :U^2, V)\big|_{p^2=-U^2} = - iU^2, \tag{4.10b}$$

$$\Gamma_o(p_1 \dots p_4, :U^2, V)\big|_{\text{s.pt. to}-U^2} = - iV, \tag{4.10c}$$

$$\Gamma_o(q/2 \; q/2 \;, -q; U^2, V)\big|_{q^2=-U^2} = 1, \tag{4.10d}$$

$$\Gamma_o(\;, q(-q); U^2, V)\big|_{q^2} = -U^2 = 0. \tag{4.10e}$$

The normalization mass U is nonintrinsic, such that its change can be compensated by a change of normalization and of coupling constant, and an additive term for $n=0$, $\ell=2$. The functional relations expressing this fact are the renormalization group equations of Gell-Mann and Low [22], the differential form of which is

$$\left[U^2 \frac{\partial}{\partial U^2} + \hat{\beta}(V) \frac{\partial}{\partial V} -2n\hat{\gamma}(V) + \ell(2\hat{\gamma}(V)+\hat{\eta}(V))\right]\Gamma_o(p_1 \dots p_{2n}, q_1 \dots q_\ell ; U^2, V)$$
$$= -i\delta_{no}\delta_{\ell 2} \hat{\kappa}(V) \tag{4.11}$$

with parametric functions with expansions analogous to (3.8a-e). Since also the Γ_{as} are the VFs of a zero-mass theory, renormalization theory [8],[9] implies that the Γ_{as} and Γ_o must be related in the analogous fashion to (3.10),

$$\Gamma_{as}(p_1 \dots p_{2n}, \; q_1 \dots q_\ell ; \; m^2, g)$$
$$= Z_1(g)^{-n} Z_2(g)^{-\ell} \Gamma_o(p_1 \dots p_{2n}, \; q_1 \dots q_\ell ; m^2, V(g))$$
$$+ i\delta_{no} \; \delta_{\ell 2} \; f(g). \tag{4.12}$$

Applying here $Op_{2n,\ell}$ of (3.7b) to both sides and using (4.6) and (4.11) yields

$$\frac{dg}{\beta(g)} = \frac{dV(g)}{\hat{\beta}(V(g))}, \tag{4.13a}$$

$$Z_1(g) = \hat{a}(V(g))a(g)^{-1} \tag{4.13b}$$

$$Z_2(g) = \hat{a}(V(g))^{-1} \hat{h}(V(g))^{-1} a(g)h(g), \tag{4.13c}$$

$$f(g) = a(g)^{-2}\, h(g)^{-2} \left[\hat{k}(V(g)) - k(V)\right] , \tag{4.13d}$$

with definition of the hatted functions analogous to (3.18). The properties analogous to (3.11),

$$Z_1(g) = 1 + O(g^2), \tag{4.14a}$$

$$Z_2(g) = 1 + O(g), \tag{4.14b}$$

$$V(g) = g + O(g^2), \tag{4.14c}$$

$$f(g) = O(1). \tag{4.14d}$$

are compatible with (4.13) only if

$$b_o = \hat{b}_o \quad b_1 = \hat{b}_1 \quad c_o = \hat{c}_o \quad h_o = \hat{h}_o \quad k_o = \hat{k}_o \tag{4.15}$$

hold, which relations, except the second, are easily verified directly and, in conjuction with (3.13), show the universality of these coefficients. In conclusion, the Γ_{as} can be computed from the Γ_o via (4.12) provided only the functions $\beta(g)$, $\gamma(g)$, $\eta(g)$, and $\kappa(g)$ are known (and an integration constant needed in $V(g) = \hat{\rho}^{-1}(\rho(g))$ is determined from a one-loop graph).

5. Exceptional momenta.

References to this section are [13] and [14].

5.1 Asymptotic forms at exceptional momenta.

Exceptional momenta were defined in Sec. 4.1 as momenta sets such that (4.2) is violated in perturbation theory, and a necessary and sufficient criterion for exceptionality at Euclidean momenta was given there If (4.2) is violated, and in perturbation theory it then is so by having terms of behaviour $\lambda^{-2}(\ell n\lambda^2)^K$ or worse, then Γ_{as} does not exist at such momenta as seen from (3.19) and (4.3). The reason for nonexistence of Γ_{as} in this case is the infrared (UR) singularity of the massless-theory VF at such momenta; in fact, the vanishing of a partial sum of momenta is a necessary (and actually also sufficient) condition for nonregularity of the massless-theory VF restricted to Euclidean momenta as proven by Ruelle [23].

For Euclidean exceptional momenta, at least the "elementary recipe" described at the end of sect. 4.2 is still applicable. If the momenta allow a partition into N even sets of zero sum each, then the leading terms as $\lambda \to 0$ are of the $\lambda^{-2a}(\ell n\lambda^2)^K$ type where $a = \max(N-2, 0)$, as a simple application of Weinberg's [21] criteria shows. The sum of these leading terms defines $\underline{\Gamma_{as}}(p_1 \ldots p_{2n}, q_1 \ldots q_\ell;\, m^2\lambda^2, g)$; the

coefficient of λ^{-2a} itself is $\Gamma_{\underline{as}}(p_1\ldots p_{2n}, q_1\ldots q_\ell; m^2,g)$. We will prove this, and obtain the analoga of (4.3) and (4.7), only for the renormalization functions at exceptional momenta; in sect. 5.3 below we consider a somewhat more general case. The nontrivial exceptional momenta sets for the renormalization functions (i.e. those with $D \geq 0$) are $(p(-p)00,)$, $(p(-p), 0)$, and $(p(-p)p'(-p'),)$.
For the first of these sets, (3.7) becomes

$$\mathrm{Op}_{4,o}\Gamma(p(-p)\ 00,;\ m^2,g) \qquad (5.1)$$

$$= -im^2\phi(g)\Gamma(p(-p)00,0;m^2,g).$$

We want to isolate on the r.h.s. the parts of behaviour $(\ell n p^2)^c$ as $-p^2\to\infty$, since it is these that violate (4.2) and have the same behaviour as the terms in the operand on the l.h.s. of (5.1). (2.17) yields (we omit arguments m^2, g where expendable)

$$\Gamma(p(-p)00,0)$$
$$= \Gamma(p(-p)\,00,)\ \Gamma(00,00) + \Gamma_{rem}(p(-p)00,\ 0) \qquad (5.2)$$

where the remainder term stems from the last two terms in (2.17) and, as discussed in sect. 2, has large- p behaviour $(p^2)^{-1}\ (\ell n p^2)^c$. Now

$$-im^2\phi(g)\Gamma(00,00) = \mathrm{Op}_{2,1}\Gamma(00,0) = \eta(g) \qquad (5.3)$$

$$\mathrm{Op}_{2,1} = \eta(g) + m^2\frac{\partial}{\partial m^2} + \beta(g)\frac{\partial}{\partial g}$$

from (2.9) and (3.7). Using (5.2) and (5.3) in (5.1) yields

$$\mathrm{Op}_{2,-1}\Gamma(p(-p)00,;m^2,g) \qquad (5.4)$$
$$= -im^2\phi(g)\Gamma_{rem}\ (p(-p)00,0;\ m^2,g)\ .$$

Omitting to write the analog of (3.19) for the present case, we immediately proceed to the analog of (4.3)

$$\Gamma_{\underline{as}}(p(-p)00,;\ m^2,g)$$

$$= \lim_{\lambda\to 0}\left[a(g)^2a(g(\lambda))^{-2}\ h(g)h(g(\lambda))^{-1}\Gamma(p(-p)00,;m^2\lambda^2,g(\lambda))\right]. \qquad (5.5)$$

Now from (3.18b) and (3.17) follows

$$h(g(\lambda))^{-1} \approx g(\lambda)^{-1/3} \approx b_o^{1/3} [\ell n\lambda^{-2}]^{1/3} \qquad (5.6)$$

such that comparison of (5.5) with (4.3) shows that

$$\Gamma_{as}(p(-p)00;m^2,g) = 0 \qquad (5.7)$$

i.e. the UR singularity for this momenta configuration is a suppression to zero. (5.7) is, using (5.5) and (5.6), sharpened to

$$\Gamma^{s(g,\lambda)}(p(-p)00,;m^2,g) \qquad (5.8)$$

$$\tilde{\sim} h(g)^{-1} b_o^{-1/3} [\ell n\lambda^{-2}]^{-1/3} \Gamma_{\underline{as}}(p(-p)00,;m^2,g)$$

for $\lambda \to 0$.

It is of interest to compare (5.8) with singularity behaviour derived directly in the zero-mass theory. The identity (2.17) we will read as an equation in Γ_{as} theory, or, for more direct reference to previous equations, in Γ_o theory related to Γ_{as} theory via (4.12). It is easy to see that the last two terms on the r.h.s. of (2.17) do exist in the massless theory by observing that wherever there is a "dangerous" propagator pair $\overset{o}{G}$, the factor to its left or right vanishes of first order at zero relative momentum thus eliminating the UR divergence in the loop integration. Since also the l.h.s. of (2.17) exists for total momentum $q \neq 0$, and so does $\overset{q}{\Gamma_o}$, $\overset{o}{I}_o \circ (\overset{o}{\Gamma}_o \circ)^{-1}$ must exist in the zero-mass theory. Moreover, since the two last terms in (2.17) actually vanish (up to logarithms) of first order as $q \to 0$,

$$\lim_{q\to 0} \overset{q}{I}{}^{o}_{o} (\overset{q}{\Gamma}_o \circ)^{-1} \equiv \underline{I}{}^{o}_{o} \qquad (5.9)$$

exists. In the Γ_{as} theory, we evaluate the l.h.s. using (4.3), obtaining

$$\lim_{q\to 0} \lim_{\lambda\to 0} \{a(g)^2 a(g(\lambda))^{-2} \Gamma((p-\tfrac{q}{2})(-p-\tfrac{q}{2})\tfrac{q}{2}\,\tfrac{q}{2},;m^2\lambda^2,g(\lambda)).$$

$$\cdot h(g)h(g(\lambda))^{-1}\, \Gamma((-\tfrac{q}{2})(-\tfrac{q}{2}),q;m^2\lambda^2,g(\lambda))^{-1} \}$$

$$= \lim_{q\to 0} \lim_{\lambda\to 0} \{a(g)^2 h(g) a(g(\lambda))^{-2}\, h(g(\lambda))^{-1}.$$

$$\cdot\ [\Gamma(p(-p)00,;m^2,g(\lambda))\Gamma(00,0;m^2\lambda^2,g(\lambda))^{-1}$$

$$+\ r(p,q;m^2\lambda^2,g(\lambda))\Gamma((-\tfrac{q}{2})(-\tfrac{q}{2}),q;m^2\lambda^2,g(\lambda))^{-1}]\}$$

$$= \Gamma_{\underline{as}}\ (p(-p)00,;m^2,g) + \lim_{q\to 0}\ \lim_{\lambda\to 0} O(q(\ell nq)^{c}(\ell n\lambda^2)^{c'}) =$$

$$= \Gamma_{\underline{as}} (p(-p)\ 00,;m^2,g) \qquad (5.10)$$

where r (.....) stands for the sum of the last two terms in (2.17) and (2.9) and (5.5) have been used.
Thus,

$$\Gamma_{as}((p-\tfrac{q}{2})(-p-\tfrac{q}{2})\ \tfrac{q}{2}\ \tfrac{q}{2}\ ,;m^2,g) \qquad (5.11)$$

$$\approx \Gamma_{\underline{as}}(p(-p)00,;m^2,g)\Gamma_{as}((-\tfrac{q}{2})(-\tfrac{q}{2}),q;m^2,g)$$

$$= \Gamma_{\underline{as}}(p(-p)00,;m^2,g)h(g)^{-1}h(g(\lambda)) \quad \Gamma_{as}((-\tfrac{q}{2})(-\tfrac{q}{2}),q;m^2\lambda^2,g(\lambda))$$

$$\approx \Gamma_{\underline{as}}(p(-p)00,;m^2,g)h(g)^{-1}\ b_o^{-1/3}(\ln \frac{m^2}{-q^2})^{-1/3}$$

where in order to obtain the last estimate we have set $\lambda^2 = (-q^2)/m^2$ and used (5.6).
The result in (5.11) is essentially identical in form with (5.8): Rather than letting the mass (by change of the bare mass only) become small at exceptional momenta, we can, with the same effect to leading order, let within the preasymptotic theory the momenta approach exceptional ones. This holds also in the other cases discussed later.
Replacing in (5.5) λ by $\lambda\lambda'$ and letting $\lambda' \to 0$ yields

$$\Gamma_{\underline{as}}(p(-p)00,;m^2,g)$$

$$= a(g)^2a(g(\lambda))^{-2}h(g)h(g(\lambda))^{-1}\ \Gamma_{\underline{as}}(p(-p)00,;m^2\lambda^2\ g(\lambda)) \qquad (5.12)$$

differing characteristically from (4.5). The considerations at the end of sect. 4.2 on the limitand in (4.3), based on (3.19), are extensible to (5.5) on the basis of the integrated form of (5.4) and lead to the validity of the "elementary recipe", as described at the beginning of this section, for the exceptional-momenta set here considered.
The homogeneous PDE for $\Gamma_{\underline{as}}$ (p(-p)00,) implied by (5.4) can also be derived using that this function is a SD expansion coefficient, see (2.17) or (5.31) below, and that the corresponding composite operator, $N_2(\phi^2)$, appearing in a Green's function gives rise merely to an additive γ-term, see (3.7), and by comparing expansions before and after mass vertex insertion. This derivation is more flexible than the one given here, which uses the explicit Green's function form of the coefficient, since the γ-terms relating to composite operators [2] are obtainable [24] using renormalization theory alone. Of particular interest for applications are light cone expansions, and by the method

just indicated homogeneous PDEs for (simple linear combinations of) the coefficient functions of the leading light-cone singularity in certain models were derived by Christ, Hasslacher, and Mueller [25]. From (2.20), in the interpretation described there, Mueller [15] has also derived a homogeneous transformation law for the asymptotic forms of the singularity-coefficient functions $\tilde{I}^{o}$.

We turn to the set (p(-p),0) and consider

$$\mathrm{Op}_{2,1}\Gamma(p(-p),0;m^2,g) = -im^2\phi(g)\Gamma(p(-p),00;m^2,g). \tag{5.13}$$

The expansion formula for the VF on the r.h.s. is (2.18), yielding

$$\Gamma(p(-p),00) = \Gamma(p(-p)00,)\Gamma(,000) + \Gamma_{rem}(p(-p),00) \tag{5.14}$$

similar to (5.2). From (3.7) follows

$$-im^2\phi(g)\ \Gamma(,000) = \mathrm{Op}_{0,2}\ \Gamma(,00) + i\kappa(g) = i\kappa(g) \tag{5.15}$$

in view of (2.13). Thus (5.13) becomes

$$\mathrm{Op}_{2,1}\Gamma(p(-p),0;m^2,g) =$$
$$= i\kappa(g)\ \Gamma(p(-p)00,;m^2,g) - im^2\phi(g)\Gamma_{rem}(p(-p),00;m^2,g). \tag{5.16}$$

Now (3.18c) implies

$$\mathrm{Op}_{0,2}\left[a(g)^{-2}h(g)^{-2}k(g)\right] = \kappa(g)$$

such that, using (5.4), (5.16) can be written

$$\mathrm{Op}_{2,1}\Big[\Gamma(p(-p),0;m^2,g) \tag{5.17}$$
$$-ia(g)^{-2}h(g)^{-2}\ k(g)\Gamma(p(-p)00,;m^2,g)\Big]$$
$$= -\ im^2\phi(g)\Big[\Gamma_{rem}(p(-p),00;m^2,g)$$
$$-\ ia(g)^{-2}h(g)^{-2}k(g)\Gamma_{rem}(p(-p)00,0;m^2,g)\Big]$$

There follows, as in earlier cases, a limit relation and the transformation formula, rewritten with the help of (5.12),

$$\Gamma_{\underline{as}}(p(-p),0;m^2,g) \tag{5.18}$$
$$= h(g)^{-1}\ h(g(\lambda))\Gamma_{\underline{as}}(p(-p),0;m^2\lambda^2,g(\lambda))$$
$$+\ ia(g)^{-2}\ h(g)^{-2}\left[k(g) - k(g(\lambda))\right]\Gamma_{\underline{as}}(p(-p),00;m^2,g)$$

equivalent to

$$Op_{2,1}\Big[\Gamma_{\underline{as}}(p(-p),0;m^2,g) - ia(g)^{-2}h(g)^{-2}\,k(g)\,\Gamma_{\underline{as}}(p(-p)00,;m^2,g)\Big] = 0, \tag{5.19}$$

or to (5.16), with Γ replaced by $\Gamma_{\underline{as}}$ and Γ_{rem} set zero. The reinterpretation of the (suppressed) mass limit relation in terms of

$$\overset{\circ}{\Gamma}_{\underline{as}} = \lim_{q\to 0}\Big[\overset{q}{\Gamma}_{as} - \overset{q}{\Pi}_{as}\,{}^{\circ}\overset{\circ}{I}_{\underline{as}}\Big] \tag{5.20}$$

on the basis of (2.18), along the lines of our treatment of (5.9), we here omit to avoid tedious repetition. We only list the analog of (5.8)

$$\Gamma^{s(g,\lambda)}(p(-p),0;m^2,g) \approx -\tfrac{1}{3}\,(\ell n\lambda^{-2})^{1/3}\,b_o^{\;1/3}\;\cdot\; a(g)^{-2}h(g)^{-2}i\Gamma_{\underline{as}}(p(-p)00,;-^2,g) \tag{5.21}$$

for $\lambda\to 0$, showing

$$\Gamma_{as}(p(-p),0;m^2,g) = \infty \tag{5.22}$$

in contrast to (5.7), and the analog of (5.9)

$$\overset{q}{\Gamma}{}^{p}_{as} = \overset{\circ}{\Gamma}{}^{p}_{\underline{as}} + \overset{q}{\Pi}_{as}\,{}^{\circ}\overset{\circ}{I}{}^{p}_{\underline{as}} + O(q\,(\ell nq)^c). \tag{5.23}$$

This formula gives a singularity at $q\to 0$ precisely analogous to the one in (5.21) because of

$$\overset{q}{\Pi}_{as} \approx -ia(g)^{-2}h(g)^{-2}\,b_o^{\;1/3}\,\Big[\ell n\,\Big(\frac{m^2}{-q^2}\Big)\Big]^{1/3} \tag{5.24}$$

as follows from (4.7) and (3.18c).

Finally, we consider the exceptional momenta set $(p(-p)p'(-p'))$, assuming for simplicity $p \neq \pm p'$. In

$$Op_{4,0}\,\Gamma(p(-p)p'(-p'),;m^2,g) = -\,im^2\phi(g)\Gamma(p(-p)p'(-p'),0;m^2,g) \tag{5.25}$$

we use (2.17) twice, obtaining

$$\Gamma(p(-p)p'(-p'),0)=\Gamma(p(-p)00,)\;\Gamma(,000)\Gamma(p'(-p')00,) + \Gamma_{rem}(p(-p)p'(-p'),0)$$

which yields, by similar steps as led to (5.19),

$$Op_{4,0}\Big[\Gamma_{\underline{as}}(p(-p)p'(-p');m^2,g) - ia(g)^{-2}h(g)^{-2}k(g)\Gamma_{\underline{as}}(p(-p)00,;m^2,g)\Gamma_{\underline{as}}(p'(-p')00,;m^2,g)\Big] = 0.$$

The analog of (5.9) is now

$$\overset{q}{I}_{as} = \overset{o}{I}_{\underline{as}} + \overset{o}{I^{o}}_{\underline{as}} \overset{q}{\Pi}_{as} \, {}^{o}\overset{o}{I}_{\underline{as}} + O(q\,(\ell nq)^{c}). \tag{5.26}$$

One can give a construction of the $\Gamma_{\underline{as}}$ functions from the zero-mass-theory VFs Γ_o without any limiting process. To this end one first defines analogs $\underline{\Gamma}_o$ of the $\Gamma_{\underline{as}}$, by using formulae analogous to (5.11), (5.20), and (5.26). E.g. $(\overline{5.9})$ now becomes using (4.12)

$$\overset{o}{\underline{I}}{}^{o}_{o} = \lim_{q\to 0} \left[\overset{q}{I}{}^{o}_{o} \, (\overset{q}{\Gamma}{}^{o}_{o})^{-1}\right] \tag{5.27}$$

$$= Z_1(g)Z_2(g)^{-1} \lim_{q\to 0} \left[\overset{q}{I}{}^{o}_{as} \, (\overset{q}{\Gamma}{}^{o}_{as})^{-1}\right]$$

$$= Z_1(g)\,Z_2(g)^{-1}\,\overset{o}{I}{}^{o}_{\underline{as}}$$

and $\overset{o}{\underline{I}}{}^{o}_{o}$ in turn is computed from (2.17) read for Γ_o VFs with q finite. Similarly, one first determines from (5.20) and (5.26), with the help of (4.12), the (finite) relation between $\overset{o}{\Gamma}_{\underline{as}}$, $\overset{o}{I}_{\underline{as}}$ and $\overset{o}{\underline{\Gamma}}_{o}$, $\overset{o}{\underline{I}}_{o}$, respectively, and limit-free formulae for the latter from (2.18) and (2.19).

In perturbation theory, one can simplify the computation of $\overset{o}{\underline{I}}{}^{o}_{o}$, $\overset{o}{\underline{\Gamma}}_{o}$, and $\overset{o}{\underline{I}}_{o}$ by exploting the "elementary recipe". (This procedure uses an idea of G.t'Hooft.) One writes the Feynman integrals for $\overset{o}{I}{}^{o}_{o}$, $\overset{o}{\Gamma}_{o}$, and $\overset{o}{I}_{o}$. The diagrams with no two-particle link $\overset{o}{G}$ in the channel carrying zero-momentum are finite and contribute in full to $\overset{o}{\underline{I}}{}^{o}_{o}$ etc. Diagrams with b $\overset{o}{G}$ - links give rise to expressions

$$\prod_{i=1}^{b} \int dk_i(-k_i^2 - i0)^{-2} \quad f(k_1 \ldots . k_b p \ldots) \tag{5.28}$$

where all integrations, except the ones related to the links, are considered carried out, and p... stands for the external momenta. $f(k_1 \ldots k_b p \ldots)$ vanishes, due to renormalization and the absolute convergence of all (Euclidean) integrals achieved thereby, for any $k_i \to \infty$. Now

$$f(k_1 \ldots k_b p \ldots) \tag{5.29a}$$

$$= \prod_{i=1}^{b} \{\left[O_{ik_i} - c(k_i,U)O_{io}\right] + c(k_i,U)O_{io}\right] \,.$$

$$. \; f(k_1 \ldots k_b p \ldots)$$

where

$$c(k_i,U) \equiv U^2\left[U^2 - k_i^2 - i0\right]^{-2} \quad , \tag{5.29b}$$

O_{ir} means to set the ith momentum r , and U is the normalization momentum in (4.10). Any lone factor O_{io} gives in (5.28) rise to an integral

$$U^2 \int dk_i^2(-k_i^2-i0)^{-2}[U^2-k_i{}^2-i0]^{-2}$$

which is discarded, leaving as contribution to $\overset{o}{\underline{I}}{}_0^0$ etc. from the diagram considered only

$$\prod_{i=1}^{b}\{\int dk_i(-k_i^2-i0)^{-2}[0_{ik_i}-c(k_i,U)0_{i0}]\}\ f(\ldots p\ldots)$$

which is UR and UV convergent. The correctness of this prescription follows from

$$U^2 \int dk\{[-(k+\tfrac{1}{2}q)^2-i0][-(-k+\tfrac{1}{2}q)^2-i0][U^2-k^2-i0]\}^{-1}$$
$$= \text{const}\ \ell n\ [U^2(-q^2-i0)^{-1}] + 0(U^{-2}q^2)$$

and the fact that the terms to be subtracted from $\overset{q}{I}{}_0^0$ in (5.27), and similarly in the Γ_0-theory analoga of (5.20) and (5.26), are, because of (4.10d-e), power series in $\ell n[U^2(-q^2-i0)^{-1}]$ without constant term.

5.2 Asymptotic behaviour near exceptional momenta.

From the fact that the set of nonexceptional-momenta sets is open, it is easily derived that in the same way as for non-exceptional momenta, the l.h.s. of (4.9) is, for $\lambda\to\infty$, approximated by the r.h.s.,

$$\Gamma((\lambda p_1+r_1)..(\lambda p_{2n}+r_{2n}),(\lambda q_1+s_1)\ldots(\lambda q_\ell+s_\ell))$$

is approximated by Γ_{as} of the same arguments, for $p_1\ldots q_\ell$ nonexceptional. However the Γ_{as} are analytic (or, if an odd partial sum of momenta vanishes, still continuous) at nonexceptional momenta, which implies that in the Γ_{as} function to the same logarithmic accuracy the r,s, momenta can be set zero, such that

$$\Gamma((\lambda p_1+r_1)\ldots(\lambda p_{2n}+r_{2n}),\ (\lambda q_1+s_1)..(\lambda q_\ell+s_\ell)) \qquad (5.30)$$
$$= \Gamma(\lambda p_1\ldots\lambda p_{2n},\ \lambda q_1\ldots\lambda q_\ell) + 0(\lambda^{3-2n-2\ell}(\ell n\lambda^2)^c)$$

if no partial sum of the p...q.. vanishes, and with a possibly less sharp bound on the difference if the p...q... are nonexceptional but an odd partial sum of momenta vanishes.
The usual power-counting argument applied to (2.17), however, yields

$$\Gamma((\lambda p+r_1)(-\lambda p+r_2)r_3r_4\ ,) \qquad (5.31)$$
$$=\Gamma(\lambda p(-\lambda p)00,)\Gamma(r_3r_4,(r_1+r_2)) + 0(\lambda^{-1}(\ell n\lambda)^c)$$

recalling (2.9). Likewise, (2.18) becomes

$$\Gamma((\lambda p+r_1)(-\lambda p+r_2),r_3) \qquad (5.32)$$
$$=\Gamma(\lambda p(-\lambda p),0) + \Gamma(,(r_1+r_2)r_3)\Gamma(\lambda p(-\lambda p)00,)+0(\lambda^{-1}(\ell n\lambda)^c)$$

and (2.19)

$$\Gamma((\lambda p+r_1)(-\lambda p+r_2)(\lambda p'+r_3)(-\lambda p'+r_4),) \tag{5.33}$$
$$= \Gamma(\lambda p(-\lambda p)\lambda p'(-\lambda p'),) + \Gamma(\lambda p(-\lambda p)00,).$$
$$\cdot\ \Gamma(\lambda p'(\lambda p')00,)\Gamma(,(r_1+r_2)(r_3+r_4)) + O(\lambda^{-1}(\ell n\lambda)^c).$$

(5.31-3) show how (5.30) is violated for particular exceptional sets of momenta p,q. In these formulae we have achieved a separation of λ-dependent terms, in which λ appears simply as scale factor to the momenta, from λ-independent terms. Obviously, in the first terms we can, to the accuracy of these formulae, replace the Γ by Γ_{as}. Consider now reading (2.17), which is an identity, for the Γ_{as} theory. Then all functions become Γ_{as} ones but the ratio $\mathring{I}^o\ (\mathring{\Gamma}^o)^{-1}$ is to be replaced by $\mathring{I}^o_{as}$. This follows from (5.5). Alternatively, we may use that all other terms in (2.17) have a $q \to o$ limit when we first replace in (2.17) everywhere total momentum zero by q (note that "zero" total momentum also enters the formula (2.7) for W) and then let $q \to o$ using (5.10). In any case, comparing the new r.h.s. with the one of (5.31) we find

$$\Gamma((\lambda p+r_1)(-\lambda p+r_2)r_3r_4,) \tag{5.34}$$
$$-\Gamma_{as}((\lambda p+r_1)(-\lambda p+r_2)r_3r_4,) = O((\ell n\lambda)^c)$$

rather than $O(\lambda^{-2}(\ell n\lambda)^c)$ as it would have been for nonexceptional momenta $p_1 \ldots p_4$ in the growing parts.- Note that for $r_3+r_4=0$, $r_3 \neq 0$ the Γ_{as} function in (5.34) would even be infinite as follows from (5.26) and (5.24). - By the same argument as just used one derives analoga of (5.34) from (5.32) and (5.33).

5.3 Application to finite-mass corrections.

We examine to what extent (4.9) can be used to determine large-momenta behaviour (in the sense of overall scaling, for the reduction of other cases to this one, see the previous sect. 5.2). The first term on the r.h.s. may be rewritten using (4.7); however, since g(λ) increases as λ increases (at least for $0<g<g_\infty$) no estimate can be obtained from (4.7) unless new, relatively speculative, assumptions are made. We defer these to sect. 6 and examine the integral in (4.9), the "mass correction" term. λ becoming large is for the last VF in (4.9) the problem of large exceptional-momenta behaviour. The analog of (4.9) for this momenta set is the integrated form of the appropriate generalization of (5.17) we first derive. We have

$$\mathrm{Op}_{2n,\ell+1}\,\Gamma(p_1\ldots p_{2n},\, q_1\ldots q_\ell\; 0;m^2,g) \tag{5.35}$$

$$= -im^2\phi(g)\;\Gamma(p_1\ldots p_{2n},\, q_1\ldots\; q_\ell\; 00;m^2,g).$$

For $p_1\ldots q_\ell$ large, the short-distance expansion analogous to (5.14) derived from (2.17) is

$$\Gamma(p_1\ldots p_{2n},\, q_1\ldots q_\ell 00) \tag{5.36a}$$

$$= \Gamma'(p_1\ldots p_{2n}\; 00,\; q_1\ldots q_\ell)\Gamma(,000)$$

$$+ \Gamma_{rem}(p_1\ldots p_{2n},\, q_1\ldots q_\ell 00)$$

with

$$\Gamma'(p_1\ldots p_{2n}\; 00,\; q_1\ldots q_\ell) = \Gamma(p_1\ldots p_{2n}\; 00, q_1\ldots q_\ell) \tag{5.36b}$$

$$+ \sum_{\text{partitions}} \Gamma(\ldots 0,\ldots)G(\Sigma'(-\Sigma'),)\;\Gamma(\ldots,\ldots)\;\ldots$$

$$\ldots G(\Sigma''(-\Sigma''),)\;\Gamma(\ldots 0,\ldots).$$

Here the sum is over all different partitions of the momenta into non-empty even sets such that the two zero-momenta are in the first and the last set; the sets are the arguments of VFs ordered in a chain, with connecting propagators transferring momenta as obvious, in a graphical picture, from momentum conservation. That (5.36) is the correct generalization of (5.14) is seen best by drawing a number of diagrams, which we leave to the reader. (It is a important to use that the momenta $p_1\ldots q_\ell$ are nonexceptional.)

We next consider Γ' and form

$$\mathrm{Op}_{2n+2,\ell}\Gamma'(p_1\ldots p_{2n}\; 00,q_1\ldots q_\ell;\; m^2,g) \tag{5.37}$$

$$= -im^2\phi(g)\{\Gamma(p_1\ldots p_{2n}\; 00,q_1\ldots q_\ell 0;\; m^2,g) + \ldots\}$$

where the dotted terms stem from the partitioned terms in (5.36b) and are in the limit under discussion negligible compared to the Γ' on the l.h.s. of (5.37) while the explicitly written VF is not. For it, we have the analog of (5.36a), derived similarly,

$$\Gamma(p_1\ldots p_{2n}\; 00,\; q_1\ldots q_\ell\; 0) \tag{5.38}$$

$$= \Gamma'(p_1\ldots p_{2n}\; 00,\; q_1\ldots q_\ell)\Gamma(00,00)$$

$$+ \Gamma_{rem}\;(p_1\ldots p_{2n}\; 00,\; q_1\ldots q_\ell\; 0).$$

Putting (5.35-38) together and using (5.3), (5.15) and the relation following (5.16) yields the generalization of (5.17)

$$\mathrm{Op}_{2n,\ell+1}\;\big[\Gamma(p_1\ldots p_{2n},\, q_1\ldots q_\ell 0;m^2,g)\; - \tag{5.39}$$

$$-ia(g)^{-2}h(g)^{-2}k(g)\Gamma'(p_1\dots p_{2n}\ 00,\ q_1\dots q_\ell;m^2,g\,]$$

$$=-im^2\phi(g)[\Gamma_{rem}(p_1\dots p_{2n},q_1\dots q_\ell\ 00;\ m^2,g)$$
$$-ia(g)^{-2}\ h(g)^{-2}k(g)\ (\Gamma_{rem}(p_1\dots p_{2n}00,q_1\dots q_\ell\ 0;m^2,g)\ +\dots)]$$

where the dots are the same terms as in (5.37), and (5.37) in the form

$$Op_{2n,\ell-1}\Gamma'(p_1\dots p_{2n}00,q_1\dots q_\ell;m^2,g) \tag{5.40}$$
$$=-im^2\phi(g)[\Gamma_{rem}(p_1\dots p_{2n}00,\ q_1\dots q_\ell 0;m^2,g)\ +\ \dots]$$

is used. (5.39) and (5.40) are the analoga of (3.7) needed now and converted by integration into the analoga of (4.9) with $\Gamma_{\underline{as}}$ functions that solve the corresponding homogeneous PDEs. There follows the generalization of (5.18)

$$\Gamma_{\underline{as}}(p_1\dots p_{2n},\ q_1\dots q_\ell 0;m^2,g) \tag{5.41}$$
$$=a(g)^{n-\ell-1}\ a(g(\lambda))^{-n+\ell+1}h(g)^{-\ell-1}\ h(g(\lambda))^{\ell+1}$$
$$\cdot\,\Gamma_{\underline{as}}\ (p_1\dots p_{2n},q_1\dots q_\ell 0;m^2\lambda^2,g(\lambda))$$

$$+i[k(g)-k(g(\lambda))]a(g)^{-2}\ h(g)^{-2}\Gamma'_{\underline{as}}(p_1\dots p_2 n\ 00,q_1\dots q_\ell;m^2,g)$$

along with the generalization of (5.12)

$$\Gamma'_{\underline{as}}(p_1\dots p_{2n}\ 00,q_1\dots q_\ell;m^2,g) \tag{5.42}$$
$$=a(g)^{n-\ell+1}\ a(g(\lambda))^{-n+\ell-1}h(g)^{-\ell+1}\ h(g(\lambda))^{\ell-1}\ .$$
$$\Gamma'_{\underline{as}}\ (p_1\dots p_{2n}\ 00,q_1\dots q_\ell;m^2\lambda^2\ ,g(\lambda)).$$

Inserting (5.41) with use of (5.42) into (4.9) yields

$$\Gamma(\lambda p_1\dots\lambda p_{2n},\ \lambda q_1\dots\lambda q_\ell;m^2,g) \tag{5.43}$$
$$=\Gamma_{as}\ (\lambda p_1\dots\lambda p_{2n},\ \lambda q_1\dots\lambda q_\ell;\ m^2,g)$$
$$-\ im^2\lambda^{2-2n-2\ell}\ a(g)^{n-\ell}a(g(\lambda))^{-n+\ell+1}h(g)^{-\ell}h(g(\lambda))^{\ell+1}\quad .$$
$$\cdot\{[\int_0^1 d\lambda'^2\phi(g(\lambda'))a(g(\lambda'))^{-1}h(g(\lambda'))^{-1}]\ .$$
$$\cdot\,[\Gamma_{\underline{as}}(p_1\dots p_{2n},\ q_1\dots q_\ell 0;m^2,g(\lambda))$$
$$-ik(g(\lambda))a(g(\lambda))^{-2}h(g(\lambda))^{-2}\Gamma'_{\underline{as}}(p_1\dots p_{2n}00,q_1\dots q_\ell;m^2,g(\lambda))]$$

$$+\ i[\int_0^1 d\lambda'^2\phi(g(\lambda'))a(g(\lambda'))^{-1}h(g(\lambda'))^{-1}k(g(\lambda'))]\,.$$

$$\cdot\left[a(g(\lambda))^{-2}h(g(\lambda))^{-2}\ \Gamma'_{\underline{as}}(p_1..p_{2n}\ 00,q_1..q_\ell;m^2,g(\lambda))\right]\}$$

$$+\ \ldots\ldots$$

where the same linear combination of VFs as in (5.39) appears. The corrections indicated by dots involve integrals over the integrals over the r.h.s. of (5.39) and (5.40). These latter can be analyzed with the help of short-distance formulae [2] extended to higher accuracy than in (5.36) and (5.38), but this need not be persued here. The case n=1, ℓ=0 of (5.43) was derived in [13], the formula for ℓ=0, n arbitrary given in [26]. Note that the complication in the computation of the "finite mass corrections" in this section stems only from the exceptional momenta, or the UR divergence of a mass vertex insertion into a massless theory involving scalar particles, which prohibit in (4.9) the trivial iterative use of the rewritten form of (4.9)

$$a(g(\lambda'))^{-n+\ell}h(g(\lambda'))^{\ell}\Gamma(p_1...p_{2n},q_1...q_\ell;m^2\lambda^{-2}\lambda'^2,g(\lambda'))$$
$$= a(g(\lambda))^{-n+\ell}h(g(\lambda))^{\ell}\Gamma_{as}(p_1...p_{2n},q_1...q_\ell;\ m^2,g(\lambda))$$
$$-im^2\lambda^{-2}\int_0^{\lambda'^2} d\lambda''^2\phi(g(\lambda''))a(g(\lambda''))^{-n+\ell}h(g(\lambda''))^{\ell}$$
$$\cdot\ \Gamma(p_1...p_{2n},\ q_1...q_\ell\ 0;\ m^2\lambda^{-2}\lambda''^2\ ,\ g(\lambda''))$$

valid only for nonexceptional momenta. In this connection cp. [27].

6. Question of true large-momenta behaviour.

References to this section are [28] and sect. IV. of [13].

6.1 Assumptions relating to fixed points.

As remarked in sect. 5.3, the $\lambda\to\infty$ behaviour of the first term on the r.h.s. of (4.9), or of the r.h.s. of (4.7), is obtainable only upon making assumptions having no basis in perturbation theory. The usual assumptions are

There exists g_∞ in $(0,+\infty)$ such that

$$\lim_{g\uparrow g_\infty}\int^g dg'\beta(g')^{-1} = +\infty, \tag{6.1}$$

and $\beta(g) > 0$ for $0 < g < g_\infty$

$$\gamma(g) \longrightarrow \gamma(g_\infty) \text{ continuously as } g\uparrow g_\infty\ , \tag{6.2}$$

and likewise for $\eta(g)$, $\kappa(g)$, $\phi(g)$

$$\Gamma_{as}(p(-p),;m^2,g)\Big|_{g\uparrow g_\infty} \text{exists for some } p^2 < 0,$$

$$\Gamma_{as}(p_1\ldots.p_4,;\ m^2,g)\Big|_{g\uparrow g_\infty} \text{exists for some Euclidean set } p_1..p_4 \qquad (6.3)$$

$$\text{and } \Gamma_{as}\ (\tfrac{1}{2}\ q\ \tfrac{1}{q}\ q,\ (-q);m^2,g)\Big|_{g\uparrow g_\infty} \text{ for some } q^2 < 0.$$

We emphasize that it is not necessary that $g_\infty<\infty$, though (6.2) and (6.3) appear less natural if $g_\infty=\infty$. The limits in (6.3) can be taken as defining renormalization conditions for a massless theory

$$\Gamma_{as}(\ldots,\ldots;m^2,g_\infty) \equiv \Gamma_{GL}(\ldots,\ldots;m^2). \qquad (6.4)$$

(6.2) gives

$$a(g)^{-1}\ a(g(\lambda)) = \lambda^{4\gamma(g_\infty)}\ e^{R(g,\lambda)} \qquad (6.5a)$$

where

$$R(g,\lambda) = 2\int_1^{\lambda^2} \lambda'^{-2}d\lambda'^2\left[\gamma(g(\lambda'))-\gamma(g_\infty)\right]\ . \qquad (6.5b)$$

Now choose $\delta(\varepsilon)$ so small that $|\gamma(g') - \gamma(g_\infty)|<\varepsilon$ for $|g'-g_\infty|<\delta(\varepsilon)$. Then

$$|R(g,\lambda)| \leq 2\left|\int_g^{g_\infty-\delta(\varepsilon)} dg'\beta(g')^{-1}\left[\gamma(g')-\gamma(g_\infty)\right]\right|$$

$$+\ 2\varepsilon\int_{g_\infty-\delta(\varepsilon)}^{g} dg'\beta(g')^{-1} + 2\varepsilon\ell n\lambda^2,\ \text{for } g(\lambda) > g_\infty - \delta(\varepsilon).$$

Herein only the last term is λ-dependent, and has arbitrarily small coefficient, such that

$$\lim_{\lambda\to\infty}\left[R(g,\lambda)(\ell n\lambda)^{-1}\right] = 0. \qquad (6.5c)$$

Similarly one proves that

$$h(g)^{-1}h(g(\lambda)) = \lambda^{2\eta(g_\infty)}e^{R'(g,\lambda)} \qquad (6.6a)$$

with

$$\lim_{\lambda\to\infty}\left[R'(g,\lambda)\ (\ell n\lambda)^{-1}\right] = 0. \qquad (6.6b)$$

Thus

$$\Gamma(\lambda p_1..\lambda p_{2n},\lambda q_1..\lambda q_\ell;m^2,g) \qquad (6.7)$$

$$= \lambda^{4-2n-2\ell-4(n-\ell)\gamma(g_\infty)+2\ell\eta(g_\infty)}$$

$$\cdot\ e^{(-n\ +\ell)R(g,\lambda)+\ell R'(g,\lambda)}\Gamma_{GL}\ (p_1..p_{2n},q_1..q_\ell;m^2)$$

$$+\ i\delta_{no}\delta_{\ell 2}\ a(g)^{-2}\ h(g)^{-2}.$$

$$\cdot\begin{cases} [4\gamma(g_\infty) + 2\eta(g_\infty)]^{-1}\, \lambda^{8\gamma(g_\infty)+4\eta(g_\infty)} \\ \quad .e^{2R(g,\lambda)+2R'(g,\lambda)}\kappa(g_\infty) & \text{if } 2\gamma(g_\infty)+\eta(g_\infty)>0 \\ (\ln\lambda^2)\kappa(g_\infty)e^{2R(g,\lambda)+2R'(g,\lambda)} & \text{if } 2\gamma(g_\infty)+\eta(g_\infty)=0 \\ [-k(g) + k(g_\infty)] & \text{if } 2\gamma(g_\infty)+\eta(g_\infty)<0 \end{cases}$$

+ (A) correction term from $g(\lambda) \neq g_\infty$ argument of Γ_{as}
+ (B) correction term from massive $\neq$ massless theory
+ (C) correction term for $\delta_{no}\delta_{\ell 2}$ part.

We note that (6.4) gives, using (4.6) (or, more precisely, (6.13) below)

$$\Gamma_{GL}(p_1\ldots p_{2n},q_1\ldots q_\ell;m^2\lambda^2) \tag{6.8}$$
$$= \lambda^{4(n-\ell)\gamma(g_\infty)-2\ell\eta(g_\infty)}\Gamma_{GL}(p_1\ldots p_{2n},q_1\ldots q_\ell;m^2)$$
$$- \delta_{no}\delta_{\ell 2}\; i\kappa(g_\infty)\tfrac{1}{4}(2\gamma(g_\infty)+\eta(g_\infty))^{-1}\;(\lambda^{4(2\gamma(g_\infty)+\eta(g_\infty))}-1)$$

i.e. the m-dependence of the Gell-Mann-Low limit theory VFs Γ_{GL} is, apart from the $\delta_{no}\delta_{\ell 2}$ term, merely the one enabling the Γ_{GL} to be scale invariant but with anomalous dimensions (see below) and to keep their mass dimension $4-2n-2\ell$.

The (B) term in (6.7) is the r.h.s. of (5.43) apart from the first term. We shall later give an argument why also Γ_{as} and Γ'_{as} occuring there should have $g(\lambda)\to g_\infty$ limits. Using (6.5) and (6.6) we find that the correction term becomes for $\lambda\to\infty$ small relative to the main term if

$$-1<2\gamma(g_\infty) + \eta(g_\infty) < 1. \tag{6.9}$$

It is obvious that the (A) part in (6.7) will become small relative to the main term if $\Gamma_{as}(\ldots,\ldots;m^2,g(\lambda))$ goes to $\Gamma_{GL}(\ldots,\ldots;m^2)$ continuously as we, of course, must assume to give meaning to Γ_{GL}. A more precise estimate can be given only if assumptions on the rate of approach are made. If one then assumes left-differentiability of $\beta(g)$ at g_∞, from (3.14), written as

$$\ln\lambda^2 = \int_g^{g(\lambda)} dg'\{\beta(g')^{-1} - (g'-g_\infty)^{-1}[\beta'(g_\infty)]^{-1}\} \tag{6.10a}$$
$$+ [-\beta'(g_\infty)]^{-1}[\ln(g_\infty-g) - \ln(g_\infty-g(\lambda))]$$

we obtain the estimate

$$g_\infty - g(\lambda) \approx f(g)\,\lambda^{2\beta'(g_\infty)} \quad , \tag{6.10b}$$

and estimates from appropriate modifications of (6.10) if $\beta'(g_\infty)=0$ but some higher derivate unequal zero, cp. [29]. (6.10b) would have to be used to estimate the (A) term.

6.2 Anomalous dimensions and conformal invariance.

(6.7) shows that

$$\dim \phi = 1 + 2\gamma(g_\infty) \tag{6.11a}$$

and

$$\dim N_2\ (\phi^2) = 2 + 4\gamma\ (g_\infty) + 2\eta(g_\infty) \tag{6.11b}$$

are the dynamical short-distance dimensions of these operators in the sense of Wilson [30]. Positivity of the metric requires

$$\gamma(g_\infty) \geqq 0 \tag{6.12a}$$

and

$$4\gamma(g_\infty) + 2\eta(g_\infty) \geqq -1. \tag{6.12b}$$

(6.11) are also the dimensions of these operators (with $N_2(\phi^2)$ appropriately redefined to avoid UR divergences) in the scale invariant GL - theory with VFs Γ_{GL}. Schroer [24] has shown that this theory, if it exists, is also conformal invariant. Conformal invariance should be helpful in determining the anomalous dimensions, though the schemes so far developed (see references in [26] and [31]) seem not to suggest yet any constructive procedure. We merely remark that also ϕ^4 theory can be written as a trilinear-coupling theory, as considered in [26] and [31], by letting $g^{1/2}\ \phi^2 = \psi$ and using a two-field Lagrangian such as

$$L = \frac{1}{2}\ \partial_\mu\phi\partial^\mu\phi + \frac{1}{24}\ \psi^2 - \frac{1}{12}\ g^{1/2}\ \phi^2\ \psi.$$

If momenta are exceptional, then (6.7) does not hold. Then the analog of the first term on the r.h.s. is to be determined from the transformation properties, like (5.41) and (5.42), of the corresponding exceptional asymptotic forms. The simplest exceptional momenta set is (p(-p 00,), and from (5.12), (6.5a) and (6.6a) we find a change of power law by a factor relative to the nonexceptional-momenta case, generic $(p_1 \ldots p_4,)$. In all other cases, there are inhomogeneous transformation laws, e.g. in (5.42) it is the linear combination of functions $\Gamma'_{\underline{as}}$ that transforms, asymptotically, apart from the $e^{R(\lambda,g)}$ factors by a power law, and in (5.39), leading to (5.41), it is another linear combination that transforms homogeneously, leading to an inhomogeneous transformation law for $\Gamma_{\underline{as}}(p_1 \ldots p_{2n},\ 0q_1 \ldots q_\ell; m^2, g)$ alone, and thus of the corresponding Γ_{GL} function.

These inhomogeneous transformation laws under scale transformation of all but the simplest Γ_{GL} lead to the appearance of logarithms if certain degeneracies in scale dimensions occur: If in

$$\{m^2 [\partial/\partial m^2] - \sum_1 \gamma\} \Gamma_{GL}(1) = \Gamma_{GL}(2)$$

$\sum_2 \gamma = \sum_1 \gamma$ holds, then the transformation law of $\Gamma_{GL}(1)$ will involve at least a logarithm; if already $\Gamma_{GL}(2)$ transforms with a logarithm, $\Gamma_{GL}(1)$ will with (logarithm)^2 etc., and we are faced with non-fully-reducible representations [32] of the dilatation group. E.G. in (5.16), in view of (5.4), degeneracy takes place if $2\gamma(g_\infty) + \eta(g_\infty) = 0$, i.e. if the dimension of $N_2(\phi^2)$ is canonical, and the logarithm is explicit in (5.18) in form of

$$\lim_{g\to g_\infty} \{a(g)^{-2}h(g)^{-2} [k(g) - k(g(\lambda))]\} \text{ prop. } \ell n\lambda$$

then, which is the same logarithm as occurs in (6.7). The degeneracy here resides in 4-dim $N_2(\phi^2)$ = dim $N_2(\phi^2)$.
Irrespective of the transformation law under scale transformation, the exceptional-momenta VFs do not transform in any simple way under conformal transformations, because the exceptional-momenta $\underline{\Gamma}_{GL}$ are extracted from the ordinary Γ_{GL} at exceptional momenta as discussed in sect. 5.1, however, the property that a partial sum of momenta be zero is not preserved under conformal transformations. Namely, the property is brought about by e.g. integrating a collection of arguments in fixed relative position over all space-time, and the integration measure over space-time changes under conformal transformations that are not Poincaré transformation. A related observation is the one in [33] that a restriction like $p^2=0$ is not preserved under conformal transformation, since the infinitesimal generators of these transformations do not commute with p^2.

6.3 Consistency check of assumptions.

It appears desirable to examine (6.3) more closely, since g_∞ is a peculiar value of the coupling constant for the Γ_{as} theory. To this end, we consider the definition of Γ_{as}, (4.9) with $\lambda=1$. It is plausible that the l.h.s. has a limit as $g\uparrow g_\infty$ since g_∞ is not a particular value of the coupling constant for the massive theory. If the Γ have limits for $g\uparrow g_\infty$, this also yields a rationale for (6.2) since the parametric functions are expressible in terms of the massive-theory VFs using (3.7) at the momenta occurring in (1.17a-c), (2.9), (2.13).
On the r.h.s., under the integral we note that (6.5b) gives

$$\lim_{g\uparrow g_\infty} R(g,\lambda) = 0$$

due to

$$|R\,(g,\lambda)|<2\varepsilon\ \ell n\lambda^2 \quad \text{if} \quad g>g_\infty-\delta(\varepsilon),$$

such that from (6.5a)

$$\lim_{g\uparrow g_\infty} \left[a(g)^{-1}a(g(\lambda))\right] = \lambda^{4\gamma(g_\infty)} \quad , \tag{6.13a}$$

and similarly (6.6) yields

$$\lim_{g\uparrow g_\infty} \left[h(g)^{-1}h(g(\lambda))\right] = \lambda^{2\eta(g_\infty)} \quad . \tag{6.13b}$$

(4.9) becomes

$$\Gamma(p_1..p_{2n},\ q_1...q_\ell;m^2,g_\infty) \tag{6.14}$$

$$= \Gamma_{as}(p_1...p_{2n},\ q_1...q_\ell;m^2,g_\infty)$$

$$-im^2 \int_0^1 d\lambda^2\lambda^{4(-n+\ell)\gamma(g_\infty)+2\ell\eta(g_\infty)}\phi(g_\infty)\ \Gamma(p_1..p_{2n},q_1..q_\ell 0;m^2\lambda^2,g_\infty).$$

Problematic is here only the convergence of the integral at 0, which leads to the large-momenta-behaviour problem for the exceptional-momenta VF we will now discuss.

To this end we only consider the behaviour of $\Gamma_{\underline{as}}(p_1..p_{2n},\ q_1..q_\ell 0;\ m^2\lambda^2,g_\infty)$ since corrections hereto lead essentially merely to repetitions of the present argument. We use (5.41) in the form

$$\Gamma_{\underline{as}}(p_1...p_{2n},q_1...q_\ell 0;m^2\lambda^2,g) \tag{6.15}$$

$$=a(g)^{n-\ell-1}a(g(\lambda^{-1}))^{-n+\ell+1}h(g)^{-\ell+1}h(g(\lambda^{-1}))^{\ell+1}$$

$$\cdot\ \{\Gamma_{\underline{as}}(p_1...p_{2n},q_1...q_\ell 0;m^2,g(\lambda^{-1}))$$

$$-i\left[k(g(\lambda^{-1})) - k(g)\right]a(g(\lambda^{-1}))^{-2}h(g(\lambda^{-1}))^{-2}$$

$$\cdot\ \Gamma'_{\underline{as}}(p_1...p_{2n}\ 00,\ q_1...q_\ell;m^2,g(\lambda^{-1}))\ \}$$

for $g\uparrow g_\infty$, $g\ (\lambda^{-1})\uparrow g_\infty$ for any fixed λ, and the argument that the $\Gamma_{\underline{as}}$ and $\Gamma'_{\underline{as}}$ in the r.h.s. hereby have limits is parallel to the one we are engaged in presently, based, however, on (5.39) and (5.40), and would lead again to the same condition (6.9) we shall find here. Indeed, taken the existence of these limits for granted, the small-λbehaviour of the r.h.s. of (6.15) is given by the one of the a,h, and k factors, which, using (6.13) and (3.18), almost cancels the λ powers in (6.14), and the convergence condition is identical with (6.9).

We conclude that if (6.9) holds, then the assumption that all Γ_{as} and

$\Gamma_{\underline{as}}$ have limits for $g \uparrow g_\infty$ is consistent, and then the "finite mass corrections" discussed in sect. 6.1 indeed are corrections only. (6.9) was derived as consistency condition in [13].

It supports this consistency argument that the $\Gamma_{\underline{as}}$ can be constructed from the Γ_{as} by formulae of sect. 2.2, as discussed in sect. 5.1, without any limit processes involved. E.g., read (2.17) for the Γ_{as} theory; in (5.10) we showed that $\mathring{I}^\circ(\mathring{\Gamma}^\circ)^{-1}$ is then to be replaced by $\mathring{I}^\circ_{\underline{as}}$. Letting then in (2.17) $g \uparrow g_\infty$ gives, for any generic momenta, $\mathring{I}^\circ_{\underline{GL}}$ expressed in terms of Γ_{GL} functions. Hereby in the construction of the subtracted B or of W no UR divergences are to be expected since (6.12a) states that the (anomalous) dimension of ϕ is larger than the canonical one! (A related fact is the absence of UR divergences in conformal invariant theory, see [34].)

7. General renormalizable theory.

7.1 PDEs for the massive theory.

Let the complete renormalizable Lagrangian involve F fields, G dimensionless coupling constants $g_1 \ldots g_G$ or (g) (i.e., coefficients of terms of operator dimension four in the Lagrangian, bosons counted with dimension 1, fermions with dimension $\frac{3}{2}$, and derivatives with dimension 1), and M(possibly generalized [30]) mass terms with coefficients $m_{1U} \ldots m_{MU}$ of mass dimension ≥ 1.

Let M_i be the ith finite (generalized) mass term, employing the appropriate normal product [2]. Then the substitution analogous to (3.1)

$$L \rightarrow L^{s_1 \ldots s_M} = L - \sum_{i=1}^{M} s_i M_i \tag{7.1}$$

results in new VFs

$$\Gamma \rightarrow \Gamma^{s_1 \ldots s_M}(p\ldots;(m),(g)) \tag{7.2}$$

$$= \prod_{j=1}^{F} Z_j\left[(s),\, (\tfrac{m}{m_1}),(g)\right]^{n_j}$$

$$\cdot\, \Gamma\{p\ldots;\left[m_1\mu_1((s),(\tfrac{m}{m_1}),(g))\right]\ldots\left[m_M\mu_M((s),(\tfrac{m}{m_1}),(g))\right],$$

$$\bar{g}_1((s),(\tfrac{m}{m_1}),(g))\ldots\bar{g}_G((s),(\tfrac{m}{m_1}),(g))\}$$

analogous to the VF-form of (3.3). Here the s_i are dimensionless, the m_i are the renormalized counterparts of the m_{iU}, $(\frac{m}{m_1})$ is the set of the M-1 mass ratios, and n_j is the number of j-type field arguments. Since, for brevity, we do not use composite-field arguments, only multiplicative renormalizations are needed. All functions Z ...,μ..., and $\bar{g}$ can be computed in renormalized perturbation theory, whereby only expansions in the factors of genuine interaction terms need be made. (See, e.g. Appendix C of reference [13], with the formula following (C.7) replaced by the correct (3.20).) Differentiating the $\Gamma^{(s)}$ with respect to the M s_i at $s_i = 0$ yields, using the Schwinger action principle [6], M inhomogeneous PDEs generalizing (3.7).

7.2 PDEs for the massless theory.

We now go to the unique asymptotic forms Γ_{as} of the Γ of the massive theory by, e.g., the elementary recipe of sect. 4.2. which here becomes: Expand $\lambda^{-D(\Gamma)}\Gamma(\lambda p\ldots;(m),(g))$ in powers of λ^{-1} and $\ell n\lambda$ for $\lambda\to\infty$, where D(Γ) is the mass dimension, equalling the superficial divergence, of Γ; the λ-independent term in this double power series is $\Gamma_{as}(p\ldots;(m),(g))$. For nonexceptional momenta, (generalized) mass vertex insertion damp the asymptotic behaviour such that

$$\Gamma_{as}^{s_1\ldots s_M}(p\ldots;(m),(g)) = \Gamma_{as}^{o\cdots o}(p\ldots;(m),(g))$$

$$\equiv \Gamma_{as}(p\ldots;(m),(g)).$$

Thus we obtain from (7.2) in obvious abbreviation

$$\Gamma_{as}(p\ldots;(m),(g)) = \Pi\, Z_j[\ldots]^{n_j}\,\Gamma_{as}(p\ldots;(m\mu),(\bar{g})) \qquad (7.3)$$

analogous to (4.5) but in different parametrization. Differentiating (7.3) with respect to the M s_i at $s_i = 0$ yields the M homogeneous PDEs corresponding to the inhomogeneous ones obtained from (7.2).
These homogeneous PDEs have the property of admitting a solution obeying (7.3), with (s)-independent l.h.s., which is in particular independent of the way one might consider the point $(s_1\ldots s_M)$ being reached from the origin in s-space in infinitesimal steps. Furthermore, in renormalized-parameter space every point can be taken as starting point with (7.3) valid. These properties imply that the M PDEs are in involution [35].
We now choose (for convenience only) the (s) such that in the massive theory all renormalized masses to generalized mass terms vanish and that the ratios of the remaining renormalized ordinary masses take fixed values, e.g. one. This then fixes M-1 of the s_i in terms of one and there remains in (7.3) only a one-parameter freedom, which is the fa-

miliar one for a massless theory due to lack of intrinsic mass scale. Expressed differently, we have formed the (in perturbation theory, unique) linear combination of PDEs that has as mass differentiation $\sum_{i=1}^{M} m_i [\partial/\partial m_i]$ which leaves mass ratios unchanged, and moreover have chosen convenient values for these ratios. The set of praeasymptotic theories Γ_{as} so restricted is isomorphic to the set of massless theories Γ_o constructed by renormalization conditions appropriate to a massless theory, analogous to the construction in sect. 4.3. These theories are parametrized, besides by the nonintrinsic overall mass scale, only by dimensionless renormalized coupling constants $V_1 \ldots V_G$ since all the renormalization conditions introducing (ordinary and generalized) masses in the massive theory are replaced by the vanishing of the corresponding Γ_o VFs at zero momenta. Indeed, would any of the generalized mass term parameters at zero momentum be nonzero, there would result UR divergences in the attempt to construct the Γ_o in perturbation theory, similar to the UR divergences one has in perturbation theory of massless ϕ^4 theory in less than four dimensions [36]
There is a relation, analogous to (4.12), between any Γ_{as} and a particular Γ_o, whereby the originally M+G free parameters are effectively reduced to 1+G as described before, the M-1 PDEs besides the overall-scaling one only describing how to convert from any mass ratios to the conveniently chosen ones.
Thus, to study large-momenta behaviour of all massive-theory VFs it suffices (in perturbation theory at least) to study the family of massless theories with VFs Γ_o. These satisfy the PDEs analogous to (4.11)

$$\{U^2 [\partial/\partial U^2] + \sum_{i=1}^{G} \hat{\beta}_i((V)) [\partial/\partial V_i] - \sum_{j=1}^{F} n_j \gamma_j((V))\} \, \Gamma_o(p\ldots;U^2,(V)) = 0 \tag{7.4}$$

with characteristics described by the G ordinary DEs

$$U^{-2}dU^2 = \hat{\beta}_1((V))^{-1} \, dV_1 = \ldots = \hat{\beta}_G((V))^{-1} dV_G . \tag{7.5}$$

These equations are the starting point of fixed point discussions such as Wilson's [28]. Computable large-momenta behaviour is obtained if there are solutions of (7.5) that lead, for $U\to\infty$, into the origin in V space. Solutions of (7.5) that lead into the origin for $U\to 0$ yield computable behaviour of appropriate massless theories in the small-momenta limit.

8. Remarks on some renormalizable theories.

This section is intended to introduce the reader to the principal facts concerning other theories than ϕ^4 and to list the relevant references without attempting completeness. Some of this material may be superseded by the time this article appears in print.

8.1 QED and gluon theory.

In QED, there is only one local mass-term insertion available, $N_3(\bar{\psi}\psi)$ in the notation of Zimmermann [2]. The PDEs analogous to (3.7), but without $N_3(\bar{\psi}\psi)$ arguments, were derived in [16]. Hereby, consistency with Ward identities as a constraint allowed to compute the γ-term related to the electromagnetic field: $e^{\ell}<(\psi\ldots\psi\bar{\psi}\ldots\bar{\psi}A_{\mu_1}\ldots A_{\mu_\ell})_+>$ and correspondingly $e^{-\ell}\Gamma(p..,q..,k_1..k_\ell;m,e)$, which are functions of e^2 only, require no γ-term for A-type arguments.
The coefficient of $\partial/\partial e^2$, if the mass term in the PDEs is $m[\partial/\partial m]$,

$$\beta(e^2) = (6\pi^2)^{-1}\, e^4 + O(e^6), \tag{8.1}$$

is gauge invariant as follows from the gauge invariance of (ordinary, and afortiori the praeasymptotic) VFs with only (≥ 4) A-type arguments. The operator $N_3(\bar{\psi}\psi)$ depends on the gauge merely through (convention dependent) overall normalization.
Only the Landau gauge, with photon propagator transverse, is stable under mass vertex insertion. Adler [29] has considered PDEs also in other gauges where, to stay in the same gauge or to have a simple gauge change, nonlocal gauge-changing insertions have to be made along with the $N_3(\bar{\psi}\psi)$ one. If the photon propagator is proportional to $g_{\mu\nu} - c\,\Box^{-1}\,\partial_\mu\partial_\nu$, with $c\,e^2$ - independent, then the gauge-change compensating term is soft (i.e. negligible asymptotically) such that the praeasymptotic Γ_{as}, or corresponding functions Γ_o, obey PDEs as simple as in Landau gauge as first observed by Wilson [28].
The theory Γ_o, see sect. 4.3, for QED is the one analyzed by Gell-Mann and Low [22]. The analog $\hat{\beta}(\hat{e}^2)$ of $\hat{\beta}(V)$, is (with mass term as for (8.1), and $z = \hat{e}^2/4\pi$, in their notation) $8\pi\psi(z)$, and has been computed to order z^3 by Rosner [37]. While the coefficients of $\hat{e}^4$ and $\hat{e}^6$ are identical (see (4.15)) to those in (8.1), a comparison of the coefficients $\hat{e}^8$ and e^8, respectively, was made by de Rafael and Rosner [38].

It is noteworthy that e, although it carries over from Γ to Γ_{as} via the preserved Ward identities, is not a conventional charge since the Γ_{as} boson propagator cannot be normalized conventionally: Since the first term in (8.1) is positive, in massless QED (i.e. in the praeasymptotic or the Γ_o theory) the small-Euclidean-momenta behaviour can be computed precisely, and one finds [20] that the boson propagator does not have a pole at $k^2=0$. Indeed, the conventional charge is zero [20] in view of the analoga in QED or gluon theory of (4.3) and (3.17).

In connection with the Gell-Mann-Low limit (in the sense of (6.4)) of QED it is tempting to use conformal invariance. However, for scale-invariant QED conformal invariance can be expected only modulo a gauge transformation [39]. (E.g., local gauge-invariance-breaking terms in the Lagrangian violate the conditions [40] for conformal invariance to be a consequence of scale invariance, since they involve derivative coupling. More generally, no Poincaré invariant gauge can be conformal invariant since for canonical vector field dimension only the longitudinal propagator is conformal invariant.) Adler [41] has proposed a class of gauges that are conformal invariant but which appear difficult to exploit. (See,e.g. [42] and references given there.) For a discussion of the eigenvalue condition for α proposed by Adler [29] see [44].

Adler [29] also considered using a finite photon mass, and this "gluon theory" has been analyzed in detail by Lowenstein and Schroer [45]. These authors in particular have used the PDEs to give a rigorous perturbation theoretical proof [46] of the Adler-Bardeen theorem [47] concerning the axial-vector-current anomaly. Gluon theory has two dimensionless parameters, e and the mass ratio, and falls under the scheme discussed in sect. 7 with M=2, G=1 (an independent $(A_\mu A^\mu)^2$ coupling being excluded as violating gauge invariance): The praeasymptotic theory is isomorphic to the one of QED.

The Ward identities are the same in the massive and the praeasymptotic theory. While the value of the bare charge is a property of the Γ_o theory alone [22], its relation to the charge e appearing in the Ward identity is in gluon theory mass-ratio-and convention dependent. There are similar mass-ratio dependences also for other asymptotic quantities as follows from sect. 7

8.2. Yukawa theories.

The parametric functions of pseudoscalar meson theory were computed to lowest order, and the corresponding eqs. (7.5) with G=2 disscussed, by

Ginzburg [48]. Zee [49] showed that for a class of simple Yukawa theories, the origin (in the two-coupling-constant space of these models) cannot be an UV stable fixed point. Coleman and Gross [18] gave an elegant proof that no renormalizable theory not involving nonabelian gauge fields can have the origin in coupling constant space as an UV fixed point. In both discussions, it was used that the renormalized coupling constant to a quartic Bose self interaction must be nonnegative, on the basis of an instability argument applicable otherwise.

8.3 Nonabelian gauge field theories.

In view of the failure of other (unobjectional, cp. [50]) models to have an UV stable fixed point at the origin (which would lead to computable large-momenta behaviour) it was of interest [51] to consider Yang-Mills theories. Politzer [52], and Gross and Wilczek [53] showed that for simple gauge groups and no, or not too many, fermions, zero-coupling is a UV stable fixed point, by computing b_0 in

$$\hat{\beta}(f^2) = b_0\, f^4 + b_1 f^6 + \dots \tag{8.2}$$

where f is the gauge coupling constant of the mass less theory and $\hat{\beta}(f^2)$ the analog of $\hat{\beta}(V)$ in sect. 4.3, and finding it to be negative under the stated condition. For SU(2), this result had already been announced by G.'t Hooft [54]. Under such circumstances, the large-Euclidean-momenta behaviour of (off-shell) VFs is characterized by fractional powers of logarithms, related to (3.17) but with $\lambda \to \infty$ rather than $\lambda \to 0$. Such models are now called "asymptotically free (in the UV)". Gross and Wilczek [53][55] also discussed the effect of Higgs-scalars on the UV stability of the origin in the (now at least two-dimensional) coupling constant space, and found that, though asymptotically free models still exist, none of them appears viable (for further discussion see [55]), and interest presently has shifted to Nambu-Jona-Lasinio type spontaneous gauge symmetry breaking (no elementary scalar fields introduced, cp. [52]) and to speculations on implications of unbroken gauge symmetries [56].

8.4 A note on asymptotically free theories.

For a reason to be explained later, it is of interest to discuss how "asymptotic freedom" (in Yang-Mills theory without scalar fields) comes about. Consider the summation of "leading logarithms" for a VF (i.e., of largest powers of $\ell n\lambda$, under overall scaling of all momenta by λ, for $\lambda\to\infty$). From sect. 4 and (8.2) we know that the characteristic quantity is

$$f^2(\lambda) = f^2\left[1-f^2 b_0 \ln\lambda^2 + O(f^2 \ln\left[f(\lambda)^2 f^{-2}\right])\right]^{-1} \tag{8.3}$$

where we have already indicated the effect of corrections.
If $b_0>0$, then for $\lambda^2 \gtrsim \exp\left[f^{-2} b_0^{-1}\right]$ $f^2(\lambda)$ becomes arbitrarily large and thus the correction term in the denominator in (8.3) arbitrarily large relative to the other terms, and (8.3) becomes inconclusive. If, however, $b_0<0$, then as $\lambda^2\to\infty$ the correction term is small relative to the dominant term in the denominator, rendering the conclusion $f^2(\lambda) \approx (-b_0)^{-1} (\ln\lambda^2)^{-1}$ reliable. The smallness of $f^2(\lambda)$ means that all graphs not contributing to leading logarithms become suppressed upon their suitable partial summation, even more strongly (though only by inverse logarithms) than the sum of leading logs itself, which is, e.g., an alternating series formally summed to (8.3) (without correction) or to certain, possibly broken, powers of that $f^2(\lambda)$.
Landau and coworkers [57] obtained the result (8.3) in QED for $e^2(m^{-2}\Lambda^2)$, which is the bare charge squared computed with a cutoff Λ, with m e.g. the electron mass. The correction term in the denominator was neglected due to the smallness of e^2 (Bogoliubov and Shirkov [8] pointed out that this is inconsistent for $\lambda\to\infty$, $b_0>0$). (8.3) was obtained by determining the graphs that would give leading logarithms and computing these logarithms by studying the integral equations obeyed by the formal sums of these graphs, which are simplified forms of the correct coupled integral equations for the Green's functions. The equations are, in QED,

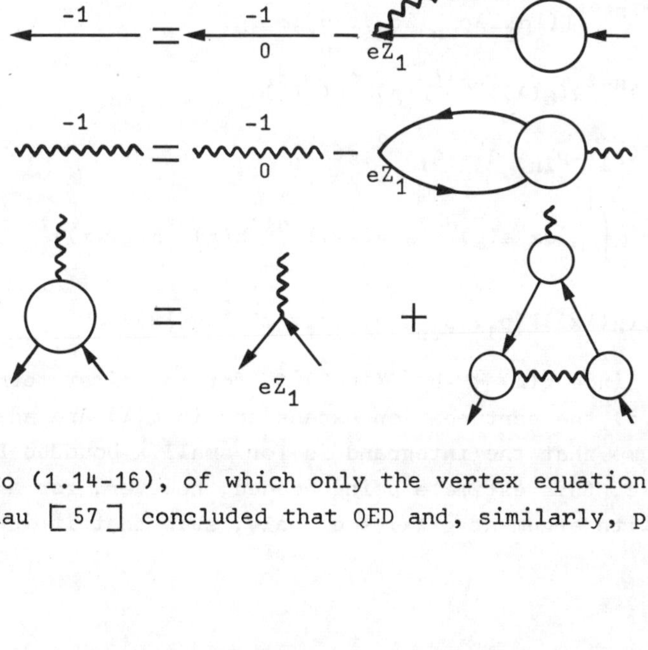

(8.4)

analogous to (1.14-16), of which only the vertex equation is approximate. Landau [57] concluded that QED and, similarly, pseudoscalar

meson theory, were theories inconsistent in the UV regime.
Now in the asymptotically free case, the existence of the praeasymptotic theory Γ_{as} <u>outside</u> of perturbation theory is less plausible than when the origin is UR stable, as in the case in ϕ^4 theory with $g>0$. Namely, the defining integral, (4.9) with $\lambda=1$, then leads at the lower integration limit to increasing coupling constant. If, e.g., $\beta(f^2)$ stays negative for all f^2 for which the theory exists up to f^2_{max}, say, and

$$\int_{f^2}^{f^2_{max}} df'^2 \beta(f'^2)^{-1} > -\infty,$$

then the praeasymptotic theory does not exists although it does, of course, in perturbation theory, however, conclusions concerning it and transcending perturbation theory, like use of the analog of (4.7) in conjuction with (8.3), are then doubtful. In such a case, one might attempt to relate the renormalization group directly to the large-momenta behaviour of Γ without the intermediary of Γ_{as}, as indicated in [50], provided the massless theory Γ_o exists; however, Γ_o and Γ_{as} are closely related. So if neither exists (outside of perturbation theory), in order to benefit from "asymptotic freedom" one could take recourse to (appropriate adaptations of) Landau's direct estimates [57], which do not presuppose explicitly the existence of the massless theory.
A rather simpler way to avoid the use of the praeasymptotic theory Γ_{as} is the following, suggested by G. Parisi: for simplicity, consider ϕ^4 theory with $g<0$, as suffices to show the method. With (2.12), (3.19) yields

$$\begin{aligned}
&\lambda^{-4+2n+2\ell}\Gamma(\lambda p_1..\lambda p_{2n},\lambda q_1..\lambda q_\ell;m^2,g) \qquad (8.5)\\
&=a(g)^{n-\ell}a(g(\lambda))^{-n+\ell}h(g)^{-\ell}h(g(\lambda))^{\ell}\\
&\cdot\ \Gamma(p_1..p_{2n},\ q_1..q_\ell;m^2,g(\lambda))\\
&+\ im^2\int_{\lambda^{-2}}^{1} dx^2 a(g)^{n-\ell}a(g(\lambda x))^{-n+\ell}h(g)^{-\ell}h(g(\lambda x))^{\ell}\\
&\cdot\ \phi(g(\lambda x))\Gamma(p_1..p_{2n},\ q_1..q_\ell;\ m^2x^2,g(\lambda x))\ .
\end{aligned}$$

For large λ, since $g(\lambda) \approx -b_o^{-1}(\ln\lambda^2)^{-1}$, for the first term on the r.h.s. of (8.5) the perturbation expansions in $g(\lambda)$ are adequate. One must now assume that the integrand is for small x bounded by $\text{const}\,x^{-2\varepsilon}$, with $\varepsilon<1$, this estimate being needed, however, for a VF and for factors a,h with argument $g(\lambda x) \leq g$ only, such that it is weaker than the condition (4.2) that implied the existence of Γ_{as}. We now split the integration region into $\lambda^{-2}\ldots\lambda^{-2\varepsilon'}$ and $\lambda^{-2\varepsilon'}\ldots 1$, with $0<\varepsilon'<1$. The first part yields $\mathrm{const}(\lambda^{-2\varepsilon'(1-\varepsilon)}-\lambda^{-2(1-\varepsilon)})$ which is negligible relative to the negative powers of logarithms one is here interested in. In the remaining part, since $\lambda x \leqq \lambda^{1-\varepsilon'} \to \infty$, perturbation expansion in $g(\lambda x)$ of the functions with this argument is adequate. A typical term is then estimated by

$$\int_{\lambda^{-2\varepsilon'}}^{1} x^{-2\varepsilon}\left[\ell n(\lambda x)\right]^{-k} dx^2$$

$$= \frac{1}{(\ell n\lambda)^k}\sum_{r=0}^{\infty}(-1)^r(\ell n\lambda)^{-r}\binom{-k}{r}\int_{\lambda^{-2\varepsilon'}}^{1} x^{-2\varepsilon}(\ell n x)^r dx^2$$

where, up to an error $O(\lambda^{-2\varepsilon'(1-\varepsilon)}(\ell n\lambda)^{\cdots})$ the integrals can be extended to lower limit zero. In the cases $n=2$, $\ell=0$ and $n=1$, $\ell=1$, the contribution from the integral in (8.5) is smaller by $(\ell n\lambda)^{-1}$ than the homogeneous part; in all other cases, it is comparable to the homogeneous part, whereby for $n=1$, $\ell=0$ (8.5) should be used upon differentiation with respect to p^2.

Closely related to these considerations is that the asymptotic expansions in inverse powers of logarithms obtained from formulae analogous to (4.5) and (3.17) satisfy generalized unitarity ([58], and second reference of [11]) in the sense of formal expansions. This is so because the Γ_{as}, whether or not they exist outside of perturbation theory, are themselves VFs of a Lagrangian field theory that satisfies in perturbation theory generalized unitarity. (Alternatively, one may consider the limit process (4.3) directly applied to the equations expressing generalized unitarity.) Some calculations hereto were given in [14] and [59], for praeasymptotic ϕ^4 theory with $g>0$ which is asymptotically free in the infrared: the same calculations but with $g<0$ then apply to large-momenta regime.

8.5 Uses of computable large-momenta behaviour.

In asymptotically free theories the large-momenta behaviour is computable, up to momenta-independent overall factors, in the deep-Euclidean region (i.e., for nonexceptional momenta) directly, and generally in the Euclidean region if appropriate exceptional-momenta techniques are used. This covers the singular coefficients in short-distance [13] and in light-cone expansions [25], (see sect. 5.1) and allows to give an explanation of Bjorken scaling as first observed by Parisi [60],

in the $(\phi_i\phi_i)^2$ theory, with g<0 however, and later proposed by Gross and Wilczek [53] (see also [61] and [62]) for asymptotically free nonabelian gauge theory for which so far no similar objection [18] as against the scalar theory is known.
The other case where asymptotic freedom has found application, e^+e^- annihilation [60][63], involves the two-current correlation function at time-like momentum, however, since this function depends on only one momentum squared and has nonnegative spectral function, the behaviour at large positive values of that invariant can relatively confidently be obtained from the one at large negative values.
Inclusive e^+e^- - annihilation has in ϕ^4 (or, more realistically, $(\phi_i\phi_i)^2$) theory been made accessible to renormalization group methods by Mueller [15] by subjecting the expansion (2.20) in the interpretation described there to the technique of sect. 5.1. It can be hoped that this process can be analyzed [64] successfully in nonabelian gauge field theory.
There is, so far, no other application known of "asymptotic freedom". The difficulty is the abundance of UR logarithms [65] in theories with vector bosons, which renders Minkowskian momenta exceptional more easily than in theories without such bosons. E.g. in (massive-photon) QED, summation of leading logarithms leads for the electric form factor of the electron to ([66][67][68], see also [69])

$$F(q^2) = \exp\left[-(16\pi^2)^{-1}\, e^2 \left(\ell n\left(\frac{-q^2}{m^2}\right)\right)^2\right] ,$$

a formula apparently unrelated to any renormalization group result since for each factor e^2 it involves a "double logarithm". It is obvious that a broad study of infrared effects is here required.

CONCLUSION

Renormalizable field theories are definable by the formal sums of perturbation series. Alternative definitions are in terms of seemingly intractable infinite systems of integral equations (e.g., [5]), or via sequences of limit procedures (finite volume, UV cutoff) possibly to be followed by difficult analytic continuations (for a presentation of these matter, see [70]) with so far, in four dimensions, uncontrollable effects.
Given the definition by series that, in at least some cases, have been shown to be divergent, the investigation of properties of the separate terms in those series is unrewarding. The remaining way then to obtain possibly relevant information is by reordering the perturbation series.

The systematic ways hereto found so far are the renormalization group, and operator product (and related) expansions. Combining these two, physical large-momenta problems can be treated insofar as they are reducible, by some expansion (e.g. by light cone expansion as in deep-inelastic e-p scattering [25]), or by highly plausible assumptions (as in e^+e^- annihilation [63]) or these two suitably combined (as in [15]) to effectively a large-Euclidean-momenta problem. Only for asymptotically free theories does one thereby arrive at predictions, precise up to momenta-independent factors. In all other cases, assumptions on the existence of fixed points [28] and determination of corresponding anomalous dimensions are still required. In the last problem, conformal invariance might be of help, whereas it is difficult to directly apply conformal invariance in situations involving exceptional momenta. A better understanding of these latter and of the infrared phenomena they are connected with is clearly desirable.

ACKNOWLEDGEMENT

A first version of these lectures was prepared and presented when the author visited the Enrico-Fermi Institute for Nuclear Studies of the University of Chicago, and he thanks professor R. Sachs and professor Y. Nambu for the invitation and for hospitality there. These lectures were also delivered at the Centro de Investigación y de Estudios Avanzados del Instituto Politécnico Nacional and at the Physics Department of Cornell University, and the author thanks professor A. Zepeda and professor D. Holcomb for hospitality at these institutions. He also thanks Ma. Eugenia López for the typing of this manuscript.

REFERENCES

1. W. Zimmermann, Comm. Math. Phys. 11, 1 (1968), 15, 208 (1969).
2. W. Zimmermann, in "Lectures on Elementary Particles and Quantum Field Theory", Eds. S. Deser, M. Grisaru, H. Pendleton, Cambridge Mass., MIT Press 1971. Annals of Physics, 77, 536, 570 (1973).
3. K. Symanzik, "Lectures on Lagrangian Quantum Field Theory", DESY T-71/1 .
   R. Stora, in "Particle Physics", Eds. C. De Witt, C.Itzykson, New York: Gordon and Breach 1973.
4. K. Symanzik, in "Cargése Lectures in Physics", Ed. J.D. Bessis, New York, Gordon & Breach 1972. (DESY 70 /62).
5. R.W. Johnson, J. Math. Phys. 11, 2161 (1970).
6. J. Schwinger, Proc. Natl. Acad. Sci. US 37, 452, 455 (1951).
7. G. Jona-Lasinio, Nuovo Cimento (L) 34, 1790 (1964).
8. N.N. Bogoliubov, D.V. Shirkov, "Introduction to the theory of Quantized Fields", New York, Interscience Publ. 1959.
9. K. Hepp, "Theorie de la Renormalisation", Berlin-Heidelberg-New York, Springer 1969.
10. T.T. Wu, Phys. Rev. 125, 1436 (1962).
11. C.G. Bollini, J.J. Giambiagi, Phys. Letters 40B, 566 (1972); G.'t Hooft and M. Veltman, CERN 73-9, Lab. I, Theor. Stud. Div.
12. K. Symanzik, J. Math. Phys. 1, 249 (1960) Appendix A.
13. K. Symanzik, Comm. Math. Phys. 23, 49 (1971).
14. K. Symanzik, Comm. Math. Phys. 34, 7 (1973).
15. A.H. Mueller, Phys. Rev. D9, 963 (1974).
16. K. Symanzik, Comm. Math. Phys. 18, 227 (1970).
17. V.V. Belokurov, D.I. Kazakov, D.V. Shirkov, A.A. Slavnov, A. A. Vladimirov, Phys.Letts. B47, 359 (1973).
18. S. Coleman and D. Gross, Phys. Rev. Lett. 31, 851 (1973).
19. J.H. Lowenstein, Comm. Math. Phys. 24, 1 (1971).
20. K. Symanzik, Springer Tracts in Modern Phys. 57, 222 (1971).
21. S. Weinberg, Phys. Rev. 118, 838 (1960).
22. M. Gell-Mann, F.E. Low, Phys. Rev. 95, 1300 (1954).
23. D. Ruelle, Nuovo Cimento 19, 356 (1961).
24. B. Schroer, Lecture Notes in Physics, Vol. 17, Eds. W. Rühl and A. Vancura, Heidelberg-Berlin-New York Springer 1973: P.K. Mitter, Phys. Rev. D7, 2927 (1973).
25. N. Christ, B. Hasslacher, A.H. Mueller, Phys. Rev. D6, 3543 (1972)
26. G. Mack, K. Symanzik, Comm. Math. Phys. 27, 247 (1972).

27. S. Weinberg, Phys. Rev. D8, 3497 (1973).
28. K.G. Wilson, Phys. Rev. D3, 1818 (1971).
29. S. Adler, Phys. Rev. D5, 3021 (1972).
30. K.G. Wilson, Phys. Rev. 179, 1499 (1969).
31. G. Mack, J. de Physique 34, Fasc. 11-12, C1-99.
32. P. Otterson, W. Zimmermann, Comm. Math. Phys. 24, 107 (1972); G.F. Dell' Antonio, Nuovo Cimento 12A, 756 (1972).
33. D. Boulware, L.S. Brown, R. Peccei, Phys. Rev. D3, 1750 (1971).
34. G. Mack, I.T. Todorov, Phys. Rev. D8, 1764 (1973).
35. E.g. E. Kamke, "Partielle Differentialgleichungen", Leipzig 1950.
36. K. Symanzik, Cargèse Lectures in Physics, 1973 (DESY 73/58).
37. J.L. Rosner, Phys. Rev. Lett. 17, 1190 (1966).
38. E De Rafael, J.L. Rosner, Ann. Phys. (N.Y.) 82, 369 (1974).
39. G. Mack, A. Salam, Ann. Phys. (N.Y.) 53, 174 (1969).
40. R. Jackiw, S. Coleman, Ann. Phys. (N.Y.) 67, 552 (1971).
41. S. Adler, Phys. Rev. D6, 3445 (1972).
42. F. Englert, J.M. Frère, P. Nicoletopoulos, Nuovo Cimento 19A, 395 (1974).
43. N. Christ, Phys. Rev. D9, 946 (1974).
44. K. Johnson, M. Baker, Phys. Rev. D8, 1110 (1973).
45. J.H. Lowenstein, W. Schroer, Phys. Rev. D6, 1553 (1972).
46. J.H. Lowenstein, B. Schroer, Phys. Rev. D7, 1929 (1973); cp. also A. Zee, Phys. Rev. Lett. 29, 1198 (1972).
47. S.L. Alder, A. Bardeen, Phys. Rev. 182, 1517 (1969).
48. I.F. Ginzburg, Sov. Phys. - Dokl. 1, 560 (1956).
49. A. Zee, Phys. Rev. D7, 3630 (1973).
50. K. Symanzik, Lett. Nuovo Cimento 6, 77 (1973).
51. K. Symanzik, in "Renormalization of Yang-Mills Fields and Applications to Particle Physics", C.N.R.S. Marseille, 72/p. 470.
52. H.D. Politzer, Phys. Rev. Lett. 30, 1346 (1973).
53. D. Gross, F. Wilczek, Phys. Rev. Lett. 30, 1343 (1973).
54. G.'t Hooft, Discussion remark at Meeting on Yang-Mills Fields, C.N.R.S. Marseille (June 1972); Nucl.Phys. B62, 444 (1973).
55. D. Gross, F. Wilczek, Phys. Rev. D8, 3633 (1973).
56. S. Weinberg, Phys.Rev. D8, 4482 (1974) and references given there.
57. L.D. Landau, A. Abrikosov, L. Halatnikov, Suppl. al Nuovo Cimento 3, 80 (1956) and references given there.
58. V. Glaser, H. Lehmann, W. Zimmermann, Nuovo Cimento 6, 1122 (1957);
K. Nishijima, Progr. Theoret. Phys. 17, 765 (1957).
59. K. Symanzik, in "Recent Developments in Mathematical Physics", Ed. P. Urban, Wien, New York: Springer 1973.

60. G. Parisi, Lett. Nuovo Cimento 7, 84 (1973).

61. D. Gross, F. Wilczek, Phys. Rev. D9, 980 (1974).

62. H. Georgi, H.D. Politzer, Phys. Rev. D9, 416 (1974).

63. T. Appelquist, H. Georgi, Phys.Rev. D8, 4000 (1973); A. Zee, Phys. Rev. D7, 3630 (1973).

64. A.H. Mueller (in preparation).

65. T. Appelquist, J.R. Primack, Phys. Rev. D1, 1144 (1970).

66. R. Jackiw, Ann. Phys. (N.Y.) 48, 292 (1968).

67. P. M. Fishbane, J. D. Sullivan, Phys.Rev. D4, 458 (1971).

68. T. Appelquist, J.R. Primack, Phys. Rev. D4, 2454 (1971).

69. V.V. Sudakov, JEPT 30, 87 (1956).

70. Lectures at Erice Summer School on Constructive Quantum Field Theory, Eds. G. Velo, A.S. Wightman (Springer 1973).

BERNI J. ALDER
Lawrence Livermore Laboratory, U.S.A.

COMPUTATIONS IN STATISTICAL MECHANICS

CONTENTS

BERNI J. ALDER
LAWRENCE LIVERMORE LABORATORY
UNIVERSITY OF CALIFORNIA

COMPUTATIONS IN STATISTICAL MECHANICS

In this series of lectures, I would like to give a survey of the enormous impact that numerical calculations have had in the field of statistical mechanics or, in other words, in the solution of the many-body problem. These calculations were made possible only through the availability of high speed digital computers and through the development of new numerical techniques uniquely suited to these computers. By these means some of the mathematical difficulties which had prevented progress in the field could be circumvented and new physical insight into the behavior of many interacting particle systems were gained. The purpose of these talks is to outline the numerical procedures without going into either excessive technical details on the computational procedures or the rigorous mathematical proofs justifying the methods used. These can be found in the references supplied. Similarly, the number of physical situations to which these numerical methods have been applied have rapidly multiplied in recent years, so that only the highlights can be presented. The topics are selected primarily on the basis of their fundamental contributions to the understanding of interesting physical phenomena and the validity of various mathematical methods to describe them, with examples chosen chiefly from my own work Further illustrations and more detail can again be found in the references, although that list is not intended, by any means, to be exhaustive. The intent of the lectures is to give a flavor of the contribution these numerical methods can make to statistical physics by discussing in a general way some outstanding examples.

I. General Considerations

Statistical mechanics is a suitable field in which to apply numerical techniques because the physics has been well established for many decades and the difficulties that are faced are of a mathematical nature. These difficulties are basically due to our inability to deal with multidimensional mathematical objects, whether they are integrals, differential equations, Green's functions or what have you. To be sure the partition function, which is the basic multidimensional quantity to be evaluated in equilibrium statistical mechanics, can be determined in

a few simple situations, such as for the perfect gas and the harmonic crystal; however, in each of these cases, the problem can be rigorously reduced to a product of one dimensional problems. Furthermore, much of the interesting physics is lost in these models.
For more realistic cases, because of the mathematical difficulties, the problem must be approximated by reducing the dimensionality of the system to low order. Most of the often used statistical mechanical theories, in fact, reduce the complex physical situation to a one particle problem, where the rest of the particles in the system merely provide an average (mean) field with which the one particle interacts. Examples of these are the van der Waals model for the thermodynamic properties and the Langevin model of Brownian motion. Another class of one particle theories might be called self-consistent models, such as the Wigner-Seitz model for a solid, the Lennard-Jones Devonshire model of a liquid, or the Hartree-Fock solution of an atom. If these models are only slightly extended to deal with two independent particles, such as in the pair product simplification used in the superposition approximation to describe fluids and the Jastrow wave function approximation to describe quantum mechanical systems, one is already faced with a formidable mathematical problem generally requiring numerical techniques for its solution.
The computer, to be sure, also can not evaluate an Avogadro number dimensional mathematical equation, which typically is required. On the other hand, the computer can be made to simulate the behavior of a many-body problem involving a few hundred particles as accurately as is required from a physical point of view. Moreover, as shall be seen, the difference between dealing with a few particle approximation as required analytically, and a few hundred particle approximation as is possible numerically is the difference between a generally poor approximation and one that almost all the time represents the properties of a macroscopic system very well.
One of the outstanding exceptions is in phase transition regions where fluctuations are large, involving more than a few hundred particles. In that case, the computer results are distorted as well, however, not so seriously that one cannot reproduce some of the analytically known results of the phase transition in the Ising model, for example. To be sure the Ising model itself represents a drastic reduction in the number of variables (degrees of freedom) with which to describe a phase transition, since the analytical solution is restricted to a two-dimensional system, the spins are confined to lattice sites and can only assume two possible values, and there must be an equal number of each type. A measure of the analytical difficulties for even this

restricted problem is that it has not yet been solved in three dimensions in spite of many tries by outstanding scientists.

(A) Limitations:

In fact, all the limitations of the numerical method are connected with the inability to deal with very large sized systems and to simulate their behavior for very long times. This is basically due to the finite memory size and limited speed of basic operations on even the fastest and biggest computers available. This situation is not likely to significantly change in the forseeable future with even more advanced computers.

The problem can perhaps best be illustrated by considering the length of time for which it is possible to follow a system of particles by the numerical method called molecular dynamics. In molecular dynamics, the classical Newtonian equations of motion of several hundred particles, all interacting simultaneously, are solved by brute force numerically by advancing the positions of the particles a small time step during which the forces on each particle can be assumed to be constant. The efficiency with which even the most advanced computer can do these time steps, however, is many orders of magnitude slower than in a real system. Thus, during a run of 10 hours on such a computer, the motion of the particles in a real liquid have been followed for only 10^{-9} sec. This precludes using this method for following rate processes having longer relaxation times than that time.

One would have thought that this would make it impossible to study any hydrodynamic phenomena since they were believed to occur on a very long time scale for macroscopically large systems. As will be seen, certain transport phenomena that were studied on computers could be identified with hydrodynamic motion occuring on a microscopic scale. Yet, the computers were limited in following this hydrodynamic motion for very long; not because, in this case, of time limitations but because the finiteness of the system caused boundary interference effects to occur. In general, to minimize boundary effects, periodic boundary conditions are imposed, but even these distort the results from infinite systems. Serious effects of boundaries are frequently investigated by studying a property for several different numbers of particles.

(B) Advantages

Another serious limitation of numerical solutions is that, like an experiment, only a specific solution for a given set of conditions is obtained. Like experiments, they can sometimes be synthesized into a general solution, which could only be recognized by having a few specific examples worked out. More frequently, a general asymptotic solution can be recognized as valid only under more restricted conditions. Most frequently approximate models can be tested for their validity and if found wanting, better approximations can be suggested which summarize the specific results. In these respects, the numerical results are often more helpful than experiments, because they can be designed for more mathematically tractable situations. Furthermore, they can give more detail than is easily experimentally accessible. For example, the potential of interaction between the particles can be made strictly pairwise additive and of a particularly simple form, such as a hard sphere potential. For such a system, the Boltzmann H-function or the triplet spacial distribution function can be evaluated to test previous hypothesis about their behavior. These are examples of functions which have not been experimentally measurable, but whose characteristic has been essential to the theoretical understanding of many particle systems. Not only is it numerically easy to evaluate these quantities, but it is possible to evaluate them over a much wider range of conditions, such as density and temperature, than can usually be covered by experiments.

The computer can out-perform experiments because it has a better resolution for small distances and short times. Most of our experimental tools have great difficulty observing phenomena at either a distance scale of the size of an atom or at the time scale of the time between collisions of two particles in a dense fluid, and that is just the scale at which the computer simulation operates. For larger distances and larger times, as pointed out before, the computer simulation fails. Since most of the statistical mechanical effects occur at the microscopic distance and time scale, the computer results have provided powerful insight into the mathematical structure of the many-body equations. By removing the restriction to tractable mathematics, it has become possible to investigate what essential physical processes dominate various phenomena. With the advantages of higher resolution than experiment it has even been possible to discover previously unsuspected physical processes.

II. Numerical Methods

There are two basic numerical procedures that have been developed for statistical mechanical investigations which would have been impossible to implement without fast digital computers. One of these has already been mentioned; molecular dynamics.[1] With that method it is possible to follow a system of particles to equilibrium and, for example, study how the Boltzmann H-function decays. In order to obtain the average behavior of the H-function, it is necessary to run a large number of systems from a series of initial states all of which are consistent with some sort of constraint. For example, for the H-function the constraint might be uniform spacial distribution of the particles (at lattice positions) over the available space but with a non-equilibrium velocity distribution such that each particle has the same speed but in a random direction.

Once the system has reached equilibrium, it is possible to determine the thermodynamic properties by taking time averages, for example, of the average kinetic and potential energy from which the temperature and the internal energy can be obtained. By this procedure, the whole apparatus of equilibrium statistical mechanics, that is the ergodic hypothesis, is by-passed. Once equilibrium is reached, it is however also possible to determine transport properties. This is done by studying the fluctuations about equilibrium, that is the transport coefficients are determined by the rate at which various fluctuations decay or dissipate.[2] Another possible way that has recently been pursued to determine transport coefficients is to set up a steady state for the particles in the presence of an appropriate external field, analogous to how such transport coefficients are traditionally experimentally determined.[3]

The other numerical method, misnamed the Monte Carlo method, utilizes the apparatus of equilibrium statistical mechanics and, hence, can only determine thermodynamic properties.[4] By comparing the results of these two different numerical procedures for identical systems for some thermodynamic property, it has been possible to get a specific instance for the validity of the ergodic hypothesis. Needless to say, the time averaged property obtained by molecular dynamics has, in the many instances that it has been checked, agreed within statistical accuracy of the result with the phase space average determined by the Monte Carlo method.

Ironically, the two numerical procedures are generally indistinguishable in terms of the amount of computer time required to obtain equilibrium properties to a given accuracy. Thus, from a numerical point of view, the formulation of phase space averaging seems unnecessary, although in some more complex situations the Monte Carlo method does have some advantage in computational efficiency. One further advantage of the Monte Carlo method should be mentioned in this connection. The molecular dynamics method is necessarily restricted to microcanonical ensemble averages due to conservation of total energy in the calculation. The Monte Carlo method, on the other hand, is much more versatile in this respect, although it is not always clear which ensemble is better in any given situation.

(A) Random Processes

A truly random (Monte Carlo) method by which to evaluate the multidimensional integral of the partition function is not practical.[5] The difficulty is that the partition function is a highly dimensional integral of a sharply peaked function; namely, the exponential of the potential energy of the system. The potential energy of the system for most of the a priori possible configurations of the molecules in the system is very large compared to those near the more probable configurations, namely those near the potential energy minimum. Thus, out of the total number of possible arrangements of molecules, only a very small fraction make any significant contribution to the integral, which is merely another way of saying the integrand is highly peaked. The Monte Carlo evaluation of a highly dimensional integral is difficult enough, but a highly peaked one presents further serious problems.

(a) Multidimensional Integration

Monte Carlo evaluation of integrals involves placing random points in the space over which the integral is to be performed, in this case it would be the volume of the system, which, since it is supposed to be infinite, presents another difficulty. For each point that is placed into the space it is ascertained whether it is under the function or outside the function. With a sufficient number of such random placing of points in this space, it is possible to determine the area under the integral since it is the fraction of the points under the function times the known total volume of the space. The assumption is that a sufficient number of Monte Carlo placements have been made to uniformly cover the available space. The accuracy with which the integral has been determined depends not only on that but also, on what might be loosely called the smoothness of the function. The density of Monte Carlo points must be so large that the average distance between these points is small compared to the narrowest region of the integrand, otherwise that region would be poorly estimated. Thus, a highly peaked function in a large space requires a very large number of Monte Carlo points. Ways to overcome that difficulty are by mathematical transformations that lead to a smoother integrand in a more limited space.

Even if that were possible, the high dimensionality would still make the problem virtually impossible, since to cover such a space even sparsely with points exceeds the limitation of computers. This can be seen from the example of covering only a ten-dimensional space with only 10 points in each dimension. The 10^{10} operations necessary, each requiring typically a minimum of 10^{-6} sec., make this problem consume one hour of machine time, without even considering the time consuming task of determining where the points fall. For this reason, practical Monte Carlo evaluation of multidimensional integrals with any accuracy is confined to less than 20 dimensions. A practical statistical mechanical application of this procedure can be found in the evaluation of higher virial coefficients.[6] Beyond the fourth virial coefficient analytical methods generally fail, even for simple model potentials, and hence the above numerical procedure can be used to evaluate a few higher ones. Beyond the eighth or so virial coefficient, the numerical procedure for the above stated reasons becomes time consuming because of the high dimensionality.

(b) Ensemble

Another form of the Monte Carlo scheme, which was actually tried, points out the same difficulty in a more physical way. Suppose you want to make up one member of an ensemble at a given number density, N/V, by randomly placing N molecules in a volume, V. After having done that, one would calculate the total potential energy of that system and according to the principles of statistical mechanics, that particular configuration would be given a probability or weight of the exponential of that potential divided by kT, that is the Boltzmann factor, where k is Boltzmann's constant and T is the given temperature. The temperature, together with the density, define the thermodynamic state of the system. This process is repeated many times to make up many members of the ensemble so that thermodynamic averages can be determined.

The difficulty with this procedure is that with overwhelming probability, a very improbable configuration of the system is selected, that is one of very high potential energy. For example, if the pair interaction potential were chosen to be that corresponding to hard spheres, almost all configurations at any but very low density would be such that some hard spheres would overlap. By overlapping is meant that the center of two hard spheres would be closer than the diameter of the hard sphere. Thus, overlapping spheres have infinite potential energy and hence the configuration has zero probability. The reason for the high probability of overlap in a randomly chosen configuration is that even in a dense gas the spheres have a somewhat ordered configuration. The higher the density, the more it is necessary to order the spheres so that they can be packed into the given volume. The greater the order of the configuration is, the less likely it will be generated from a random selection of positions.

(c) Wrong Ensemble

To overcome this difficulty, one might have thought one could place the particles into the volume one at a time.[7] Thus, a sphere would be placed randomly into the volume only if that sphere did not overlap with any of the spheres previously placed. If it did overlap, it would be rejected and another tried. This procedure to build up a member of an ensemble has two serious flaws. One is a practical one, namely, it is still not possible to reach very high density. At a density approximately corresponding to the one where a real system would be at its critical density, it is no longer possible to place another particle into the system, that is, virtually every placement is rejected. This is again because random placements lead to inefficient packing with large spaces between the spheres which cannot accomodate another sphere after a rather low density is exceeded. The other flaw is a fundamental one, namely the method does not select a proper member of an ensemble. The only statistical mechanically sound procedure is to select all the positions at one time from a set of random numbers, as was first suggested, and then reject or accept the configuration of spheres according to whether there are any overlaps. The procedure of putting the spheres in one at a time selects a set among the random numbers which leads to wrong weighting of the members of an ensemble. Perhaps the simplest way to illustrate that is to show that for three hard spheres the probability of obtaining a given final configuration is different if they are placed in one at a time depending on the order in which they are placed, thus invalidating the procedure. The probability of placing a sphere into the volume is proportional to the volume accessible. As a consequence, if the final configuration of the three spheres is such that two of them are so close that a third sphere cannot be placed between them, then the probability of placing a third one far away from these two is different than if that third one had been accepted first and one of the close in ones last.

(B) Monte Carlo Method

Since the random procedures fail, it is necessary to go to the next order of stochastic complexity to evaluate statistical mechanical averages, and that is by means of a random walk in configuration space.[8] That random walk generates a Markov chain in which each state depends only on the previous state. Furthermore, the transition probability between two successive states must be such that averages over all the states developed in the Markov chain correspond to an ensemble average. The problem is to determine these transition probabilities to achieve this aim, namely the one that each state generated recurs with a frequency corresponding to its Boltzmann weighting. If configurations could be preselected with Boltzmann probability, then Boltzmann averages can be obtained by weighting all the generated configurations equally. This process corresponds to importance sampling where only configurations near the energy minimum are generated thus eliminating the wasteful procedure in random sampling where many improbable configurations have to be discarded.

The transition probability is not hard to find as illustrated for a dense system of hard spheres. A number of spheres are initially placed into the volume in any non-overlapping configuration, usually lattice positions, according to the density to be investigated. Next, a sphere is selected randomly and displaced by a random amount in a random direction. The size of the displacement is in principle irrelevant; however, in practice it determines the rate at which the system wanders through phase space. This is easily visualized since too small displacements leave the system little changed for many such displacements, while too large displacements have the same effect since the displacements are rejected. This is because after an attempted random displacement of a randomly chosen particle, the displacement is only accepted if it does not lead to an overlap with any sphere in the system. If it does not lead to overlap, another member of the Markov chain has been generated. If overlapping results, the move is rejected, and the particle replaced, and the old configuration must be counted again in any averaging procedure over the Markov chain. The latter requirement must be imposed so that the system satisfies microscopic reversibility. More loosely speaking, the configuration must be recounted because the rejection of the move indicates the higher probability of the original configuration.

(a) Ergodicity

A succession of such random displacements of randomly chosen particles then constitutes a random walk in phase space. The only other requirement on the transition probability besides microscopic reversibility, and the trivial one that the probability is normalized, is that it is possible to go from any accessible state to any other by a series of finite one particle displacements; that is, that the system is ergodic. That has more to do with the choice of the initial state of the system and the number of attempted moves, than with the choice of the form of the transition probability. One defines a system as being ergodic and in the same class if by a series of single moves one can reach any state of the system independent of the initial state.[9]

In proving ergodicity for the Markov chain under consideration, one can take advantage of the fact that on the computer one is necessarily dealing with a finite chain, because the position of the particles can only be specified in a discreet space since there is an upper limit to the significant numbers by which the positions of particles can be specified. Hence, the Markov chain is necessarily ergodic in the sense that there is a mean finite recurrence time for each state in the same class. More troublesome is whether the finite time is sufficiently short in terms of the number of moves required, so that reasonable computer runs can cover phase space sufficiently well. There is no theoretical guidance on this point and usually no difficulty is encountered as judged empirically by averages over runs of varying length. Some practical difficulties are, however, encountered, as will be seen, when there are two separate or nearly separate regions in phase space. If there are separate regions in phase space then only averages over each class of such states is possible and the class is entirely determined by the choice of initial states. An example is the glassy and crystalline state of hard spheres at high density. It is possible to so jam up a finite number of hard spheres in a given volume that there is no way to rearrange them into the thermodynamically more stable crystalline state. This is a situation where for a finite system configuration space has been separated into two classes. For an infinite system, there is always an improbably large density fluctuation which will allow the glass to rearrange itself into the crystalline form. In that case the two classes of configuration in phase space are connected by a narrow passage, reflecting the improbability of that event. Thus, one has to face the problem that for finite systems phase space can be distorted and that rigorous proofs of ergodicity for such systems are likely to be confined to a given class.[10]

A case where different classes of phase space are connected by a narrow passage occurs also for hard spheres in the fluid-solid phase transition region; the two classes corresponding to the fluid and solid pockets. As will be seen, the finite system jumps sufficiently rarely from one pocket to the other that it has not been possible at some densities to get a meaningful average over both pockets for even the longest computer runs on the fastest machine available. As a consequence, the average over each pocket has to be given separately as if the thermodynamic functions at these densities were two valued.

(b) General Procedure

The impression that these problems only occur for hard sphere systems should be dispelled. For a more realistic interaction potential between a pair of particles, the repulsive forces still predominate at sufficiently high density leading to similar difficulties. For such a realistic potential the random walk procedure must be slightly modified. This time a random displacement of a randomly chosen particle is allowed with Boltzmann probability by evaluating the energy change resulting from the move. The move is allowed with probability of the exponential of this energy change divided by kT. This is technically achieved by comparing a random number with that exponential. If the random number is less than the exponential, the move is allowed, if not, the move is disallowed and the configuration obtained by replacing the particle at its original position is counted again in the statistical mechanical averaging.
Starting from some arbitrarily chosen initial configuration, most moves initially will be allowed as the system moves towards its energy minimum. After the system is near that state, it moves away from that minimum only with Boltzmann probability. It is only then that the averaging procedure is instituted; the early part, as the system moves toward equilibrium, being discarded. This initial phase is usually over very quickly. After equilibrium is reached, some unfavorable moves must be allowed, otherwise eventually the system would get frozen into its lowest potential energy state; that is all possible moves would increase the potential energy. Under those circumstances it is no longer possible to take averages.

(C) Molecular Dynamics

In comparison to the Monte Carlo method, the molecular dynamic method is completely deterministic.[11] The particles are given positions and velocities and the equations of motion are followed. This can be done more accurately and faster if the potential of interaction is chosen to be of the hard sphere or square-well type. For such potentials, the only forces are impulsive ones at the discontinuities. The system of simultaneous differential equations then reduces to a system of simultaneous algebraic equations which can be solved for the time to the next collision among all the set of particles. Proceeding collision by collision the time evolution of the system is obtained. There are some obvious technical arrangements of the calculation that make this procedure efficient when a large number of particles are followed. In the Monte Carlo calculation, as well as in molecular dynamics, the calculation can be arranged such that in a given amount of computing time the number of collisions calculated is independent of the size of the system. For more realistic, continuous potentials this is also true. There one assumes that the forces are constant over some short time interval after which the forces are recalculated. The length of the time interval over which it is legitimate to assume constancy of the forces depends on the accuracy with which one wants to solve the equations. There are standard predictor-corrector methods with which these equations are solved which allow determination of the appropriate time interval.[12]

(a) Accuracy

One must make sure that the physical relaxation process one wishes to study has a shorter relaxation time than the time scale of build-up of numerical error. This is not always easy since it is amazing how rapidly numerical imprecisions accumulate. A simple demonstration of this can be obtained by running a system forward for a certain time and upon reversing all velocities ascertain whether the system returns to its original state when it is run again for the same time. Even with twelve digit arithmetic for the simple hard sphere system, the time corresponds to only about 10,000 collisions, pretty independent of the size of the system. Different round-off errors lead eventually to a different collision history. One might argue that that should not affect the physical result since real systems are subject to random external effects as well; however, to be sure that the physical result is not affected, it is wise to run some calculations with higher accuracy, such as double precision, to make sure that the numerical noise does not dominate the longer physical rate processes that one studies.

The situation is particularly bad in stellar evolution calculations[13] where most of the time the stars are weakly interacting through the gravitational potential and rarely strongly interacting through a collision. Furthermore, the system needs to be followed for a very long time. The accumulation of numerical errors is so severe (exponential growth) that most calculations have been dominated by numerical problems, putting the physical results in doubt.

(D) Thermodynamic Properties

In molecular dynamics, the energy as well as some other conserved quantities are continuously monitored to check on gross computer errors as well as round off errors. There is, hence, no problem in obtaining the internal energy. The pressure can be obtained through direct application of the virial theorem since the forces are known at each instant of time or through an indirect application of the virial theorem, namely through averaging by means of the pair radial distribution function. Although knowledge of the internal energy and the pressure over the entire density and temperature regime completely determines all the thermodynamic functions of the system, it would be highly desirable to have a direct measurement of the other thermodynamic properties at a given state. Oddly enough, the numerical schemes can directly determine thermodynamic properties which are derivatives of the partition function, but not the thermodynamic properties which depend on the partition function itself, without some additional calculations. Thus, it is possible to get the Helmholtz free energy by integrating the pressure with respect to volume from infinite volume to the desired volume, but not by calculating it by means of the logarithm of the partition function, since the latter is unobtainable directly.

(a) Free Energy

A number of special techniques have been developed to circumvent this problem, particularly for the free energy or chemical potential since that quantity is very important in a number of applications. All these techniques depend fundamentally on the fact that the chemical potential is the derivative of the Helmholtz free energy w.r.t. the number of particles. Hence, the probability of adding an additional particle to the system is just the thermodynamic activity, whose logarithm is proportional to the chemical potential.[14] The idea was used a long time ago in calculating the free energy of a system of charged particles by slowing turning on the charge of one particle. In general, integration over a "coupling constant" is called for, whose function it is to turn whatever potential one is studying on, all the way to full strength. In this way the free energy is identified with the reversible work necessary to add an additional particle to the system. Although this still calls for a series of computer runs for various strengths of the interaction, with a subsequent integration over this strength parameter, it sometimes is preferable over the volume integration called for by the direct thermodynamic formulation.

The technique of determining the probability of acceptance of an additional particle is preferable whenever it is possible to do it. As pointed out earlier, that technique fails at higher density because the probability of acceptance becomes too low. This probability is determined by periodically stopping a molecular dynamics run for hard spheres, for example, and trying via many random tries to insert another sphere somewhere into the volume of the system. The probability of insertion is just the ratio of successful insertions, namely those that do not overlap with the other spheres in the system, to the total number of tries. In general, the activity is the average of the exponential of the energy change relative to kT caused by the insertion. Since at high density the overwhelming number of insertions lead to very unfavorable energy changes, it is not possible to obtain a proper measure of this probability and some other process, such as the coupling process, must be used.

Later, use will be made of another coupling process where each particle in the system is confined to a single cell which is then allowed to get larger and larger. In this way the free energy of a solid can be obtained by integration over the size of the cell. Such a molecular dynamics run would have hard sphere particles collide either with other particles as before or with the walls of the confining cell.

(E) Quantum Systems

So far only numerical procedures for the solution of the classical many-body problem have been discussed. The numerical procedures to solve quantum statistical mechanical problems are in a very much more rudimentary stage, and the problem is so complex that it taxes even the most advanced computers by presently available schemes. To permit the equilibrium quantum many-body problems to be solved by a probabilistic method, the Schroedinger equation must be transformed into the form of a diffusion equation. This is easily done by transformation to an integral equation incorporating the relevant boundary conditions. That integral equation involves an unknown Green's function which must be first generated by a Monte Carlo game, after which the diffusion-like equation can be solved by another Monte Carlo game.[15] This procedure has been successfully used to determine the properties of a few hundred Boson particles at zero degrees. Extensions to higher temperatures and Fermi statistics involve further enormous practical difficulties.

To study the more spectacular transport properties of quantum systems, the time dependent Schroedinger equation must be solved. This might be carried out via the so-called Madelung transformation which converts the Schroedinger equation to a Navier-Stokes type hydrodynamic equation, one for each particle in the system.[16] One is then faced with the horrendously difficult task of the solution of a large number of coupled Navier-Stokes equations. In as much as these procedures are still in a very rudimentary stage with no physical insight of general interest so far forth-coming, the subject will not be pursued further here.

III. Phase Transitions

One of the first problems to which these numerical techniques were applied was to the question of a solid-fluid phase transition for hard spheres. There had been some theoretical work suggesting such a transition, but of dubious reliability because of the approximations introduced. The significance of the proof of such a melting transition lies in whether purely repulsive forces suffice, as had been suggested by more physically based models. Furthermore, if such a transition exists, the problem would have been reduced to its simplest and essential characteristic, giving hope that a rigorous theoretical proof would be forthcoming.

Theories of phase transitions represent one of the great challenges in statistical mechanics. As mentioned before, only the Ising model has so far yielded to that challenge. The difficulty is that the partition function has a mathematical singularity at the conditions of a phase transition beyond which the nature of the solution changes. There have been many suggestions of divergence of asymptotic series representations of the partition function at the conditions of the phase transition. One of these is the divergence of the virial expansion which represents the partition function in a power series in the density. The numerical results can shed some light on whether this density series might diverge and how reliable Padé approximants are in extrapolating such a series.

(A) Fluid Density

In as much as the first few virial coefficients were known, the hard sphere system was first run at sufficiently low density where the known virial series converged in order to check the numerical procedure. Once satisfied, higher virial coefficients were numerically obtained and the equation of state (pressure-volume-temperature relation) was extended to higher densities by the Monte Carlo and molecular dynamic techniques.[17] One of the important facts that was learned in these calculations was that once more than 100 particles were used, the results no longer depended significantly on the number of particles; that is, the results did not change by more than about 1%, which is the typical numerical uncertainty in the results. Some studies were carried out to theoretically predict this number dependence, but in general, that turns out to be a more difficult problem than to solve the actual infinite particle problem in the first place. Hence, the important conclusion which makes all these numerical simulation calculations so significant, namely that 100 particles represent the properties of an infinite system well in almost all situations, is based on such empirical observations.

The virial series represents the fluid equation of state well below a certain density. Beyond that density, using Padé approximants to extend the series proved to be a hazardous procedure. The problem is that there exist several Padé approximants by which to extrapolate and there is no theoretical guidance as to which one is best. These several Padé approximants all satisfy the general feature that they are the ratio of two polynomials which, when divided, reproduce the known virial coefficients. They differ in the degree of the polynomial in the numerator and denominator. However, all of them gave no indication of any discontinuity in the equation of state. Instead, they seem to follow continuously from the fluid phase into a glassy phase. This glassy phase is generated on the computer by slowly compressing the fluid phase, leading eventually to one of those isolated pockets in phase space spoken of earlier. For this "glassy" phase the pressure goes continuously to infinity as the system approaches a state of what might be called random close-packing.

Comparing the fluid properties with the prediction of the superposition approximation theory led to the conclusion that that approximation fails rather badly at high density and thus its veracity in the prediction of a phase transition must be questioned. On the other hand, the Percus-Yevick or scaled particle theory, proved to be much more accurate and since the results can be expressed in simple analytical form, that approximation has become very popular, although the reason for its validity remain obscure and how to extend the theory has proved difficult. The hypernetted chain theory proved to be of intermediate accuracy compared to the above two theories. It is an interesting sidelight that, in general, the numerical work required to solve the above theories is of comparable magnitude to that for the numerically exact simulation procedures and this, usually, without assurance of a unique solution.

(B) Intermediate Density

At intermediate densities a peculiar behavior reminiscent of a phase transition was observed when the system was started initia-ly in a solid or crystalline phase. After a very long computer run where the pressure fluctuated a small amount around the solid pressure value, the pressure suddenly jumped and fluctuated a small amount around the extension of the fluid branch of the equation of state. Only rarely did the pressure jump back down again. It was not possible to properly average over these two branches since transitions were too rare. The two branches were identified by various ways as fluid and solid by such measurements as the diffusion coefficients and, perhaps, most convincingly by a graphical display of the trajectories of the particles. These trajectories showed the localized excursion of the particels in the solid around their lattice positions and the accessibility of all of configuration space to the particles in the fluid phase. This two branched behavior only occurred for systems of greater than about 30 particles.[18] For fewer particles, the system stayed in the solid branch even though it could be determined that particles were able to exchange lattice positions. The point seems to be that there were not enough particles present to create a disordered state, since after such an exchange in a few particle system the particles found themselves again in an ordered state. The lesson to be learned is that to obtain even the rudiments of a phase transition it is necessary to deal with a fairly large number of degrees of freedom. The only way that can be done analytically is by the introduction of collective coordinates, as will be illustrated shortly.

Going to much larger systems than 100 particles proved fruitless in three dimensions. Systems as large as 4000 particles have failed to show the coexistence of the two phases. However, in two dimensions, that is for hard disks, the effort was successful.[19] Oscilloscope traces of the trajectories showed the coexistence of solid and fluid regions, and the equation of state in that region showed large pressure fluctuations whose average connected the solid and fluid branches by a van der Waals-like loop. Although for infinite systems it can be proven that such a loop cannot exist, for finite systems it is possible.

The loop was analyzed in terms of the contribution of interfacial tension, which causes the predominant phase to force the particles in between the two different regions to take on more of its characteristics.[20] In this way, it is possible for a finite system to have a region which would be, for infinite systems, mechanically unstable; namely one where the pressure goes up as the volume goes up.

(C) Communal Entropy

The numerical hard sphere results allowed for the first time an evaluation of the so-called communal entropy that dispelled some of the misconceptions previously held about it.[21] The communal entropy concept arises from the difference between the properties of a system when it is treated in the one particle approximation and when it is treated as a many-body system; in other words, it is the quantity one was able only to speculate about previously. One thing that is known about it, however, is that in the high density (close-packed) limit of a hard sphere solid, the one particle concept is rigorously valid.[22] On the other hand, in the low density limit the one particle concept leads to an error in the entropy, S/Nk, of unity; the error arises from the difference of a particle being confined to a region of a volume per particle as opposed to having the entire volume of the system accessible. This difference in the entropy, the communal entropy, must appear somewhere between the high and low density limit, and the logical place appeared to be at melting, when indeed the particles in the solid, confined to the neighborhood of a lattice site, are released upon melting from that restriction. The argument was considerably strengthened by the empirical observation that melting entropies for simple substances were of that magnitude. A detailed analysis revealed, however, that the communal entropy upon melting was only a small fraction of unity and that the communal entropy appeared gradually over the entire region from solid to gas.

The communal entropy concept could be utilized in two further ways. One of these was to predict the shift of the observed phase transition with the number of particles.[20] The nagging worry of the doubting Thomases about this hard sphere melting transition continues to be whether this phase transition disappears for infinite systems. It is hence necessary to understand how the transition depends upon the number of particles. Besides the decrease of the amplitude of the van der Waals like loop as the number of particles increases, as already discussed, the transition shifts in pressure and in volume difference between solid and fluid. The latter can be investigated but will not be pursued here. The pressure shift can be understood on the basis of the communal entropy. This arises because the finite system of N particles in a periodic cell can be considered as differing in entropy from an infinite system by the communal entropy of N particles per cell. The other use of the communal entropy was to establish the coexisting conditions for the solid and fluid phase in three dimensions, where sufficiently big systems could not be investigated on the computer to establish these conditions directly.[22] What is hence necessary is to establish the free energy of the solid to be equated with the known free energy of the fluid. The way to accomplish that is by extending the solid branch of the hard sphere system artificially to low density by imposing the restriction of single occupancy. Single occupancy leads to a continuous extension of the solid branch to low density which upon volume integration of the pressure leads to a free energy all along the real solid region.

(D) Mechanism

There is little doubt that the phenomena observed on the computer corresponds indeed to melting. Using more realistic potentials than hard spheres, the phase transition observed on the computer could be well correlated with actual data. Furthermore, the phenomena could be understood on the basis of a well established empirical model of melting, namely Lindemann's law. That law, which gives a very good account of melting in all sorts of materials, states that a substance will melt one its temperature and density are such that the root mean square displacement of a particle from its lattice site is about 10% of the lattice spacing. Both the hard sphere and disk systems correspond well to that rule. The problem now is to develop a model or a mechanism which accounts for this observation.

The safe and pragmatic view is that a substance melts once its free energy equals that of a liquid. Proposals for an instability in the solid associated with melting usually lead to an extension of the solid branch into the metastable region, or a van der Waals loop coexistence curve. The problem with instability theory hence is, that one has to find the first instability mode which, if not identical to the phase transition condition, lies at least closest to it. Furthermore, instability theories cannot identify the new phase formed. However, the universality of melting leads one to believe that there is a mechanism, namely one that destroys the long range order by an instability to a long wave length shear mode. Although there have been some arguments that this cannot hold for disks since there is no long range order, the arguments are deceptive. To be sure, it can be shown that no long range order exists in a two dimensional solid because the mean square displacement is infinite. This is because the system acts like a jelly. However, there is long range order relative to the position of any particle.

The search for a simple mechanism for melting led to a soft mode instability which can be mocked up crudely in a correlated cell model. In the two dimensional version of that model, alternate rows of atoms move collectively relative to each other. At the point where one row of atoms can slip past the neighboring ones, the model leads to a discontinuity in the equation of state represented by a distorted van der Waals-like loop. The conditions at which the transition occurs in the model correspond nearly quantitatively to the ones found for disks.

A search for evidence that rows of atoms slip past each other in molecular dynamics calculations in the solid phase of disks very close to melting proved successful. The sliding rows consisted typically of 5 atoms each in two neighboring rows, where the end atoms jumped across the rows to close the loop. The loops got bigger closer to melting until finally one end no longer closed the loop. That created a defect in the solid which propagated very rapidly, leading to melting. There exists also some experimental evidence in the body-centered cubic phase of helium just prior to melting, that the heat capacity has a small contribution due to this collective mode. Usually that contribution to the heat capacity is unobservably small, but in helium a number of factors make it more favorable to observe it.

(E) Lattice Models

No theory has as yet been able to prove the existence of the melting transition for hard spheres. Such a proof remains one of the outstanding challenges in equilibrium statistical mechanics. The question one might then ask is whether it is possible to prove such a transition for lattice models, such as was used in the Ising model. For lattice models, such a proof might be easier because, as pointed out before, they involve a reduced space. In this case, the continuous configuration space would be replaced by a discrete space, or to state the problem mathematically, the integration over volume in the partition function is replaced by a sum. Replacing an integral by a discrete sum is done all the time in numerical integration and as the mesh or grid is made sufficiently fine, the approximation to the integral can be made as accurate as desired. The analytical problem, however, is that for fine meshes the integral can no longer be managed. The only lattice problems that have been solved correspond to mesh points comparable to the number of particles in the system. In other words, unless a grid spacing comparable to the size of the particle suffices, the evaluation cannot be handled analytically. The unfortunate fact is that such coarse spacing seriously distorts the partition function in dense fluids and hence the phase transition can not be adequately described by practically workable lattice models.

A one-dimensional model using hard rods can demonstrate the limitations very simply, since its partition function can be worked out exactly, no matter how fine the grid.[25] At typical melting conditions, the spacing between particles, as Lindemann predicts, is about 10% of the size of a particle. Under these conditions, in order to evaluate the partition function accurate to 1%, that 10% space must be covered by 50 grid points or there must be 450 grid points to cover the space occupied by each particle. In three dimensions that number must be cubed, and with that mesh size only 1% accuracy is achieved, and even then it is not assured that the narrow passages in phase space between the fluid and solid pockets might not be missed. At the critical point, where the lattice theories have been much applied, the situation is not nearly as bad. At the critical point the spacing between particles is comparable to the size of the particle itself. In order to get 1% accurate thermodynamic properties, it would be necessary to cover a particle with only 50 points in each dimension, still way beyond analytical capabilities. However, by covering each particle and each of the spaces in between particles by at least one grid point, as is typically done, one might hope to get at least the qualitative behavior correct, particularly for the very important long range behavior of the spacial correlation functions. These correlations, being of a range much larger than the size of a particle, should not be qualitatively dependent on the grid size as long as it is also of the size of a particle.

The lattice models, nevertheless, have found two useful applications in dense systems. One is that they can provide both upper and lower bounds for the partition function. The bounds for hard rods derive from the simple observation that if one measures the non-integral length of a line by a discrete number of equal sized elementary lengths that fit in it, one always overestimates the length. The average error due to edge effects is just one half of the length of one of these elementary length, and it is this fact which led to the error estimates. The other practical application is that no matter how fine the grid, the lattice model gives the wrong behavior in the close-packed limit, where the space between the particles shrinks to zero. Thus, the computer calculations with their exceedingly fine grids, determined by the number of significant digits with which the positions are specified, cannot properly determine the properties of dense systems very near close-packing. This has not been a serious limitation, and, in fact, through the computers it has been possible to establish the correct asymptotic behavior of hard sphere solids.[22] This will be discussed in connection with critical behavior in solid systems.

IV. Van der Waals Model

One of the results of the numerical work has been to gain more insight into the validity of the mean field models of which the van der Waals model is a prime example. The van der Waals model, which is now just over a hundred years old, was neglected by theorists for a long time during which they grappled with many unphysical models as they tried to cut through, mostly unsuccessfully, the mathematical difficulties by approximations that would make the mathematics tractable. These efforts ignored the mounting empirical evidence that the van der Waals model correlated the experimental data remarkably well outside the critical region. The numerical work circumvented these mathematical difficulties, showed the validity of the van der Waals model in the asymptotic high temperature limit for particles with a hard core interaction potential, showed that the corrections to the model were small, and showed that all three phases of matter could be obtained by that model.

(A) Perturbation Theory

As a matter of fact, the usefulness of the van der Waals model and its corrections from a numerical point of view transcends its usefulness as a physical model. From a numerical point of view, it is highly desirable to overcome the limitations of numerical calculations that they are specific to all the conditions imposed in the solution. One of these conditions is the choice of the specific interaction potential between a pair of particles, for example, the hard sphere potential. It would be very valuable if this specific solution could be extended by analytical means to a class of slightly different interaction potentials, for example, those which have a weak attractive potential added on to the core. It turns out that if this attractive potential is sufficiently weak so that first order perturbation theory suffices, and sufficiently long range so that this perturbation theory can be extended over the entire density regime, one recovers rigorously the van der Waals model.[26] Such a perturbation theory will not only lead to the gas-liquid coexistence region for which it was originally proposed, but also to the solid-fluid phase transition region, since, as has just been seen, this is contained in the hard sphere system by itself. Without the addition of attractive forces, the hard sphere fluid cannot distinguish between a liquid and a gas phase.

Having a rigorous proof on the conditions under which the van der Waals model is valid, it is now of interest to ascertain what the corrections to the model are when these conditions are not met, since for many realistic potentials the conditions on the range and strength of the attractive potential are far from being met. The limitations imposed by the short range of many realistic attractive potentials can be overcome by restricting the density range over which the van der Waals model is applicable. At sufficiently high density (for practical purposes usually greater than the critical density of a substance) the mean space between particles is sufficiently small compared to the range of the attractive forces, so that the particles rarely escape the interaction potential of their neighbors, which, in effect, provides the mean potential. One must investigate why, in spite the large value of parameter of the perturbation expansion (the strength of the attractive potential relative to kT) the model gives accurate results when only the first term, that is the van der Waals model, is used. For typical liquid conditions the parameter of the potential relative to kT is of the order of unity and hence the precision with which the van der Waals model describes the liquid state means that the higher order corrections must be small.

(B) Pure System

There is a well founded statistical mechanical theory for each coefficient of the power series expansion in the perturbation parameter.[27] The latter is the reciprocal of the reduced temperature; that is the temperature reduced by the strength of the attractive potential. For evaluation of each successive coefficient in this expansion, higher order correlation functions of the unperturbed system must be known, or equivalently, higher order fluctuations in the attractive potential about the mean (more precisely cumulants) must be evaluated.[28] The latter approach is numerically much more convenient since it avoids having to tabulate highly dimensional correlation functions for the unperturbed system. As is typical of perturbation theory, each succes-

sive term requires more detailed information about the unperturbed system, which, in this case, must be numerically obtained.
As expected, even the first correction to the van der Waals model made only a small contribution. This is the square term in the reciprocal temperature, involving quadratic fluctuations about the mean potential. The reason that the still higher order terms are negligibly small was found to be due to the near Gaussian character of the fluctuations. For a Gaussian distribution the higher order cumulants can be shown to be strictly zero. The half width of the Gaussian distribution, needed for evaluation of the first correction to the van der Waals model, was found to be remarkably insensitive to density over the entire fluid range; corresponding to a root mean square fluctuation of about half a particle around the mean of about twelve particles at normal liquid densities. At lower densities the mean number of particles within the range of the attractive potential decreases to about three particles near the critical density of a fluid; however the fluctuation about this lower mean is still about half a particle. The reason for the relative larger fluctuations at the lower density is, of course, the looser packing of the hard spheres. It is, hence, entirely a matter of the geometric behavior of hard spheres that leads to the near volume independence of the Gaussian half width for typical ranges of the attractive forces. The significance of the volume independence is that it makes that quadratic correction term to the van der Waals model negligible for the equation of state. Since the still higher order terms were negligible in the first place, the validity of the mean field approximation, even for strong attractive forces, can be understood.
Only near the critical point, where the fluctuations are large, does this expansion not converge. The hard sphere system has no premonition of such large fluctuations and hence it is not surprising that it does not even give the qualitatively correct singular behavior of the critical point. Neither does a finite system of particles interacting via a realistic potential, when studied by the Monte Carlo or molecular dynamics method. The critical singularities are distorted by any calculation that constrains the long range fluctuations. Nevertheless, these calculations or the perturbation calculation can locate the critical point occurrence as to pressure, volume and temperature very well.

(C) Mixtures

There were several interesting questions regarding mixtures that could

be pursued by numerical calculations. One of these concerns the possibility of the existence of two fluid phases for mixtures of hard spheres. This could be ruled out on the basis of these calculations.[29] In fact, it was found that these mixtures of spheres behaved nearly ideally. Their non-ideality could be identified with a small negative excess volume of mixing, that is the volume of the mixtures was slightly smaller than the sum of the two volumes of the pure hard spheres systems that are mixed. This corresponds to the elementary observation that it is possible to pack mixtures of spheres slightly more efficiently than the pure spheres separately. The near ideality of these sphere mixtures does not correspond at all to what is found in real mixtures, so that we conclude that the attractive forces play an essential role in determining the excess properties of mixtures.
Hence one is led to an application of the van der Waals model to mixtures. Though that model does very well in the prediction of the qualitative features of mixture properties, as shall be discussed in the next section for gas-gas phase separation, it leaves something to be desired in its quantitative aspects. The reason for the greater inadequacies of the van der Waals model for mixtures than for pure systems is readily apparent. In the first place, the excess properties, representing a difference, are generally an order of magnitude smaller than the pure properties, thus higher precision is required to predict them well. Secondly, deviations from the mean field theory are not only due to potential fluctuations as in pure fluids, but also due to concentration fluctuations. These have been ignored or were poorly represented in previous mixture theories that tried to improve upon the van der Waals model. Testimony to that fact is that no previous theory gave significantly better results than the van der Waals model. That model necessarily assumes random mixing; whereas in real mixtures, particularly when phase separation is approached, significant clustering must occur. This clustering, corresponding to concentration fluctuations, can be accounted for by calculating the corrections to the mean field model in the perturbation series.
Although that series can account for the major quantitative defect in the van der Waals mixture theory, the convergence of the series is not as fast as one would like. This shows up particularly in the heat capacity, which is entirely dependent on the fluctuations; the van der Waals model predicting zero excess heat capacity. In fact, due to large concentration fluctuations, the heat capacity in a mixture is frequently larger than the sum of the separate heat capacities of the pure components at the same density and temperature as the mixture.[30]

The first correction to the van der Waals model, the quadratic term in the fluctuations, cannot account for this phenomena, so one is forced to go to higher order in the series. Instead of performing these rather time consuming numerical calculations, an effort was made to reorder the perturbation series in powers of the reciprocal temperature by graph theoretical techniques so as to achieve faster convergence.[31] In this way, certain classes of graphs occurring in each term of the invinite reciprocal temperature series was summed, such as ring graphs, for example. The evaluation of these graphs does not require more knowledge about the unperturbed systems than is already necessary for evaluation of the mean field. The disadvantage of this method is that one loses the simple power series behavior in temperature and that one no longer has a simple parameter characterizing the expansion. The latter difficulty is related to the well known problem of ordering sets of graphs according to the magnitude of their contribution. One can, however, use physical arguments to show why the summed graphs make the dominant contributions and the omitted ones are of lower order. The test of such an approach lies in its predictive power and judging by the calculation of the heat capacity of mixtures, the approach proved very successful. Thus, the numerical approach was able to pinpoint the type of correction required to the van der Waals model of mixtures, and a relatively simple graph theoretical calculation was able to account for it, making deviations from random mixing calculable.

V. Systems with Critical Behavior

As was mentioned before, the numerical schemes dealing with a finite number of particles seriously distort the nature of the singularity at the critical point which crucially depends on the large distance behavior of correlations. Thus, it cannot be expected that the exponents signifying the nature of the singularity at the critical point can be realistically obtained. One might, however, again ask how large the system has to be in order to approximate the macroscopic system well with respect to the onset of the singularity close to the critical point. In other words, is it possible to observe deviations from classical (van der Waals) behavior close to the critical point for the very small systems that can be investigated on computers. Sufficiently far from the critical point, the fluctuations might be small enough so that these finite systems might realistically represent them and yet the fluctuations are not so small that the mean field theory predictions are accurate.

Very little work has been done in this connection because of the enormous computer time requirements to get sufficiently precise results. A very preliminary investigation shows promise that such an investigation might be fruitful for some properties.[38] The results indicated that the pressure-volume-temperature relations on the coexistence curve around the critical point had too small a deviation from classical behavior to be observed within the accuracy of the data. However, the heat capacity did show enhanced values in the critical region that are inconsistent with mean field theory predictions. Extracting an exponent from this numerical data in a fashion analogous to the way real data is handled, led to a divergence consistent with what is experimentally observed. The conclusion appears to be that small scale fluctuations significantly contribute to the heat capacity some distance away from the critical point and they can be quantitatively evaluated in small systems. Closer to the critical point, the heat capacity for these finite systems does not assume the large values found experimentally, again because the large scale fluctuations that are ignored have a large effect.

(A) Solid-Fluid

The computer methods have been applied in a number of cases to the question of whether a critical point exists, ignoring the nature of the singularity. One of these examples concerns the question of the existence of a solid-fluid critical point. The previous work on the hard sphere solid-fluid phase transition showed that if the intermolecular repulsive potential is sufficiently repulsive, there is no such critical point since it was found that at sufficiently high density, no matter what the temperature, a solid phase is formed. Since real systems have softer repulsive potentials than hard spheres, the question is then whether for these a solid can always be formed. The hard sphere calculations were, hence, repeated for the repulsive Coulombic potential, which represents the interaction potential between atoms at very high pressure, once the electronic shell structure has been pressure ionized.[32] The electrons can then be considered as forming a uniform background in which the charged nuclei swim. These nuclei, because of their heavy mass, behave classically and interact via the softest known repulsive potential, namely the Coulombic one.
This potential, because of its long range nature, must be handled with some care in finite systems with periodic boundary conditions. It can be shown that better results are obtained at high density, if one abandons periodic boundary conditions and instead considers the finite system as constituting a unit cell, periodically repeated, to form an

infinitely large crystal. The practical difference is that a particle not only interacts with all the particles in the cell but also all their periodic images. The contribution of the periodic images is evaluated analogously to the methods used in evaluating potential contributions in ionic crystals, namely by the Ewald lattice sum method. The use of the more complicated boundary conditions slows the calculation down, but not prohibitively.

The results were that again a solid is formed when the potential energy of the system exceeds a certain multiple of the kinetic energy. That multiple is not of the order of unity as it is with most ordinary materials, but about a factor of one hundred larger. Nevertheless, the melting transition is still well described by the Lindemann law in the sense that melting occurs when the root mean square excursion distance from a lattice site is about twenty per cent of the lattice spacing. This is to be compared to the ten per cent value for more repulsive potentials.

(B) Coulomb Gas

Although these results lead to the conclusion that no solid-fluid critical point exists, one must, at sufficiently high density, consider that the nuclei start to behave quantum mechanically. At these very high densities, it can easily be shown that the zero point energy must eventually exceed the lattice potential energy of a Coulombically interacting system. This is because the zero point energy rises as the reciprocal square of the lattice spacing while the potential energy rises only as the reciprocal lattice spacing itslef. Thus, there is a density even at 0°K above which a solid cannot exist. This density corresponds to a lattice spacing of about 0.1 Bohr radius for protons and is obtained by an approximate numerical calculation, since the full quantum mechanical many-body solution procedure is as yet too cumbersome to be practical.[33] The scheme involves approximating the total wave function as a pair product wave function of a specified functional form. The form involves a few parameters, which have to be optimized variationally. This can be achieved by a series of classical Monte Carlo calculations with the introduction of a fictitious interaction potential related to the specified pair wave functions. This approximation has been shown to be quite adequate in explaining the melting transition in helium.[34]

The conditions under which the Coulomb gas has been found to melt, again corresponds quite well to Lindemann's law, except that in this

case, the particles oscillate about their lattice position due to zero point energy instead of thermal energy. A melting curve for the quantum Coulomb gas can be obtained at other than 0°K by utilizing Lindemann's law, where the excursion distance is a result of a combination of zero point and thermal motion. This leads to a closed region in density and temperature below which a solid exists. The surprising prediction of this theory is that for protons above about 3000° K it is not possible to have a solid. That temperature is very low by astrophysical standards and indicates that most protonic stars have liquid interiors. This does not appear to be true for pulsar for which star quakes seem to be observed. Hence, either the potential or the density do not conform to the above conditions in these stars.
It should be emphasized that the limited region over which a solid is stable in a Coulomb gas does not imply the existence of a critical point as it is usually understood, namely an existence of a phase transition which is less than first order. It can be argued that such a transition would inplausibly imply that one could continuously go from a solid of long range order to a fluid of short range order. The above calculations, in fact, predict a first order solid-fluid transition.

(C) Helium

Another prediction, very similar to the above, of a closed solid region of stability is in the case of helium. In this case, the solid region is surrounded by another solid phase and again the transition is characterized as first order and a temperature above which the specified crystal structure cannot exist.[35] The reason for taking this system up briefly here is to point out that such a phase transition is extremely hard to study numerically because the difference between the two crystal structures is so extremely small. The two crystal structures, face-centered-cubic and hexagonal - close-packed are so similar that numerical methods have not yet established their relative stability. A determined effort was made to establish for classically behaving hard spheres which of the two phases would be stable as a function of density.[36] The free energy difference could not be reliably established. Even at high density where special methods related to the single occupancy calculation were used to calculate the coefficients of the expansion in the free volume, the first two coefficients proved to be the same within 5 significant numbers. The third coefficient, but not outside statistical error, favors face-centered cubic. For a harmonic oscillator it can be shown by still other numerical methods peculiar to that potential, namely by the solution of the Born-van Karman matrix for the frequency modes, that the face-centered crystal is stable.[37]

The above calculations of the stability of the face-centered cubic phase in the high temperature classical region agrees with experimental observations on helium. The stability of the hexagonal crystal structure of helium at low temperature is no doubt due to the quantum mechanical nature of these crystals. In fact, it can be rigorously proven in the single occupancy approximation for hard spheres, by making use merely of the symmetry of the crystal, that at low temperature the hexagonal phase is more stable than the face-centered phase.[38] Thus, hard spheres are likely to exist in at least two different crystal structures, but to establish the phase transitions by numerical techniques requires higher precision methods than are presently available for even the simpler classical calculations. In this connection it is worthwhile to point out that hard spheres are not stable in the body-centered phase, at least classically. This phase is also known to exist for helium, but its stability in a classical system requires at least non-nearest neighbor forces. In confirmation, a body-centered crystal of hard spheres partially melts at any solid density when investigated by numerical means.

Another interesting point emerged from the study of hard spheres at high density. This is a value of the entropy at close-packing, which is a measure of the free volume available to a sphere.[39] It turns out that the self-consistent model makes an enormous error in that estimate. A much more accurate value is obtained by the so-called Lennard-Jones Devonshire model which allows a central sphere to have access to the entire volume of a cell formed by holding the nearest neighbors fixed at their lattice positions. The moral appears to be that self-consistency is not always advantageous from a quantitative point of view. Disappointing too was the slow convergence of various solid cluster theories that tried to estimate this free volume more accurately by calculating corrections to the single particle cell models.

(D) Electronic

The numerical calculations can also help to elucidate certain electronic transitions which are known to have real critical points in the solid phase. Such a critical point is believed to be possible only in the solid phase if the transformation is between two identical crystal structures. The only known example, namely cerium, conforms to that. The transformation is believed to be caused by the stability of a different electronic configuration of the atom in the two phases. The change of electronic configuration with pressure comes about because of the shift of electronic energy levels when the atom is confined to a smaller space. Confinements favor those energy levels where the wave function has a higher probability to be, so to speak, inside the atom. For present purposes, such a sudden shift in electron configuration can be idealized as a sudden shift in the radius of an atom. Thus, a numerical study with a two step hard sphere interaction potential, that is a repulsive square well potential, might mock up this physical situation well.

Although little work has been carried out for such a potential as yet, it can be predicted that if the change of radius is less than 10%, the system might show a critical point. This is because, recalling Lindemann's law, a solid system can "average" over such distances before it melts. If the radius change is larger than 10%, then one would predict that the solid-solid phase transition extends all the way to melting. Upon melting, however, the phase change would disappear since a liquid can "average" over such distances. One would, in fact, predict a melting maximum in the neighborhood of the solid and melting phase line intersection. There is one known example of this, namely cesium, which is likely to have a very large size change in its electronic transition. In any case, the numerical calculations with such potentials could confirm the prediction that no first order phase change would be observed in liquids. There is no known observation of one.[40]

(E) Gas-Gas

For mixtures, the critical point of pure systems becomes a critical line in the pressure-temperature plane. Gas-gas critical behavior is then said to occur if this critical line reaches temperatures greater than the critical temperature of either pure component. The conditions under which such behavior can occur has been worked out long ago through the van der Waals equation. Experimental confirmation of this phenomenon is also quite old. The criterion for which this effect will be observed requires that the mixture consists of two components with largely different attractive forces. With the revival of interest in the critical point of mixtures, particularly tricritical points, it is worthwhile to point out that the numerical procedures can also be utilized to make more accurate predictions of critical lines and their intersection than the van der Waals predictions.[41]

VI. Transport Properties

(A) Previous Models

The first numerical effort in the study of transport properties went into critical tests of existing models. One of these models proposed that transport in fluids was primarily of the type occuring in solids with vacancies. In such a solid a particle would typically oscillate for long periods of time around its lattice position and occasionally, when it is near a vacancy, make a jump of the order of a lattice spacing into the vacancy position. A critical test of this mechanism appears to be a study of the distribution of free paths in a fluid. The free path in a hard sphere fluid is the distance a sphere traverses between successive collisions. If the above model were applicable, one would expect this free path distribution to be bimodal. This is because one would expect two characteristic distances to be of importance, namely the small distance corresponding to the free length and the one corresponding to the average spacing between centers of particles. Instead, the free path distribution showed no significant structure.[42] It proved to be nearly exponential with distance and nearly independent of density when scaled by the mean free path. Thus, from this evidence, there appears to be no significant new mechanism of transport in dense fluids. The simple kinetic theory valid at low density describes the free path distribution well at all densities. Another model proposed that the trajectory of a typical particle consisted of two unrelated parts, namely hard sphere collisions that could be described by kinetic theory and interaction with the attractive potential in between the hard core collisions that could be described by Brownian motion. The use of Brownian motion was thought to be applicable because the interaction with the attractive potential was thought to be weak and frequent. A numerical study of the frequency of interaction with the attractive interaction in a square-well system revealed that such interactions were rare at fluid densities.[42] In a typical liquid, the majority of successive hard sphere collisions were found to be uninterrupted by an attractive interaction. In fact, this study led to the conclusion that the essential characteristic of transport in liquids could be understood if transport coefficients for hard spheres could be calculated. This is in conformity with the accuracy of the mean field approximation for equilibrium properties. In the mean field approximation, the attractive potential does not affect the transport properties at all, since a constant potential cannot exert a force and, hence, affect the trajectory of a particle. These preliminary numerical studies, hence, proved the linear or free flight trajectory model to be the most appropriate. Hence, a detailed study of hard sphere transport properties was undertaken.

(B) Comparison to Experiments

Just to make sure of the validity of the hard sphere model, a comparison with experiments was undertaken using an approximate hard sphere transport theory.[43] That theory, developed by Enskog, assumes that the transport coefficients calculated at low density, where the Boltzmann equation is assumed to be a valid description, can be scaled to higher density by merely scaling the time between collisions. Just as the free path distribution was found to be accurately scalable by the mean free path, so it is assumed that if there is no new mechanism of transport with density, the transport coefficients need only be scaled in time owing to the fact that the collision rate increases with density. Since the collision rate can be directly related via the virial theorem to the equation of state previously determined, the comparison to experiment on such substances as the rare gases could be easily carried out. The result for the diffusion coefficient, the viscosity and the thermal conductivity was in agreement within about 10% in the dense fluid region. Applicability of the model, just as in equilibrium, is restricted to the dense fluid region and, in fact, the single parameter necessary for this experimental comparison, namely, the hard sphere diameter, was obtained from thermodynamic data analyzed by means of the van der Waals model. By using the somewhat empirical scheme of making the hard sphere diameter temperature dependent to account for the softness of real repulsive potentials, the temperature dependence of the transport coefficients could also be well explained without invoking the customary device of an activation energy. It is this latter concept which had been disproved by the free path study.

(C) Perturbation Theory

The conclusion from this study is that the correlations in the hard sphere systems, omitted in the Enskog theory, make a small contribution to the transport coefficients. The small deviations that were observed in the experimental comparison are certainly also partially caused by the neglect of the attractive forces in the hard sphere model. Thus, it is not only necessary to get accurate properties for hard spheres by molecular dynamic computer studies, but to develop a perturbation theory for the attractive potential, just as for equilibrium properties. The latter theory has met severe obstacles. The basic difficulty is that a small potential perturbation can in certain circumstances lead to vast changes in the trajectory of a particle. For example, such a small perturbation can, by bending the path of one particle a small amount, cause two hard spheres to collide which would not have otherwise, changing the whole subsequent history of the system. Another example is that a small potential perturbation can lead to orbiting in what would have been without such a perturbation a glancing encounter. Although some individual trajectories can thus be vastly changed, it is known from the above experimental comparison, that on the average, the attractive potential has a relatively minor effect. The difficulty appears in the theory through divergences in the perturbation expansion when it is applied directly to the individual trajectories. It is, hence, necessary to perform the averaging over the trajectories first, before carrying out the attractive potential perturbation. When that is done, the expansion series is no longer a simple power series in reciprocal temperature.[30] From a numerical point of view, it has not been possible to get an explicit expression for the coefficients in the perturbation expansion, or in fact, the form of the expansion. Instead, one is forced to the cumbersome device of calculating the transport coefficients by molecular dynamics at several high temperature values, for example, for a square-well potential, and from that data by differencing schemes extract the expansion coefficients.

(D) Method

There are two basic approaches by which to determine transport coefficients by the numerical schemes. One is analogous to the typical classical experiment whereby an external gradient is imposed.[3] Since this necessarily involves imposing steep gradients on such small systems so that a flow can be determined outside statistical fluctuations, these calculations are not favored. Other disadvantages are that frequently artificial boundary conditions must be imposed and that only the transport coefficients themselves, and not the detailed correlations underlying them, can be obtained. This, coupled with the fact that these calculations are not even computationally significantly faster, has led to the use of the fluctuation-dissipation approach. This approach is also increasingly being used to determine the transport coefficients experimentally.

Taking the simplest transport coefficient as an example, the diffusion

coefficient is defined as the long time limit of the mean square displacement of a particle in an infinitely large system divided by the time. It has always been cautioned by the careful investigator that this limit will only exist, that is that the diffusion coefficient is a constant, if the particle forgets its past history sufficiently fast The usual argument presupposes that after a sufficiently long time has elapsed, the particle assumes a stationary Markov process, that is it executes the random walk of a drunkard, for which this limit can be proven to exist. There are, of course, well known examples where this does not occur. One is the perfect gas in which a particle never collides, so that the distance it travels is proportional to the time elapsed. Hence, the diffusion coefficient diverges linearly in time. At the other extreme is the particle bound by a harmonic oscillator. In that case, the particle is confined to a finite region in space, and the diffusion coefficient vanishes in the long time limit.
The above expression for the diffusion coefficient is called the Einstein expression and there exist equivalent ones for all the other transport coefficients and these have, in fact, been programmed on the computers.[47] A single trajectory in equilibrium phase space is run, and the mean of the square of the displacement for the diffusion coefficient, for example, is determined by chopping this single trajectory up into a number of equal segments. Each initial state of these segments can be considered as another member of an ensemble. Alternatively, averaging can be achieved by generating members of an ensemble by the Monte Carlo method whose time evolution is subsequently followe by molecular dynamics.

(E) Autocorrelation Function

Although a rigorous proof of the identification of the measured transport coefficients with the theoretical Einstein expression is lacking, there is no difficulty showing the equivalence of the Einstein expression to an autocorrelation function expression. By straight forward mathematical manipulations one can show the equivalence of the mean square displacements over time to the time integral of the velocity autocorrelation function, provided the diffusion coefficient itself exists. This equivalence has also been shown numerically and sometimes one or the other expression is preferred for technical reasons. For discussion purposes, it is preferable to speak of these autocorrelation functions, because they show most clearly the effect of correlation. More precisely, the deviations of these autocorrelation functions from the exponential behavior predicted by the Enskog model of uncorrelated events, reveal what the dominant correlations are that have to be accounted for in a more accurate theory of the transport coefficients of hard spheres.

Very little is rigorously known about these autocorrelation functions. The Enskog theory predicts rigorously only the initial slope since for very short times no correlations can build up. The next term in the Taylor expansion in time of the autocorrelation function already involves ternary collisions which could be correlated through such cyclic events as particle 1 colliding with 2, 2 with 3 and subsequently 3 with 1 again. This contribution, which Enskog neglects, is already hard to calculate. The numerical work confirmed the very early time behavior as predicted by the Enskog theory, and showed, that the relaxation time determined by it, when introduced into an exponential autocorrelation function, reproduced the full autocorrelation function fairly well. Although the deviations from exponential behavior found by molecular dynamics were small, they were of a puzzling nature at intermediate densities.[45] At high densities, the velocity autocorrelation function became negative, as expected, due to the solid-like character of the fluid, which causes the volocity of a particle, on the average, to be reversed after a few collisions. At intermediate densities, however, the velocity correlations were positive and lasted for many collisions. A typical particle seemed to persist in its original direction of motion after having undergone many collisions. To be sure, the persistence was quite weak, at a level of about one hundredth of its original velocity, but such that it led to a significant dependence of the result on the number of particles used in the molecular dynamic computations. This number dependence, which for more than one hundred particles is usually hardly detectable, as well as the long nature of the persistence indicated that a highly collective effect was involved.

(F) Hydrodynamic Model

After some false starts, this persistence was eventually quantitatively explained as being due to a hydrodynamic vortex mode.[46] The hydrodynamic model that was used involved a moving volume element, representing the particle, in an otherwise stationary viscous, compressible medium, representing all the other particles of the system. The moving volume element pushes the fluid ahead of it and leaves a smaller than average pressure behind it. The pressure equalized by the generation of a compressional sound wave ahead of the volume element and a dilational one behind it. The speed with which this sound wave propagates depends on the compressibility, but in any case does not lead to any persistence. The pressure can, however, also equalize by the generation of a double vortex flow around the moving volume element. At long times, this flow dominates since the sound speed by simple dimensional arguments can be shown to propagate faster than the diffusion of vorticity.

The vortex flow leads to positive persistence, since some of the momentum transferred to the medium is fed back into the particle itself. On a molecular scale, the particle collides with the neighbor ahead of it, which subsequently collides with the other neighbors such that some of the original velocity is returned. One then expects that the maximum positive persistence should appear when approximately 10 collisions, corresponding to the number of nearest neighbors at liquid densities, have occurred. The numerical results verify this expectation and furthermore show that at lower densities the maximum persistence appears at fewer collisions, till at very low densities, the persistence is so weak that it cannot be observed. The clinching numerical verification of these ideas was a determination of the velocity field itself.[47] This was determined by a study of the velocity correlations between a particle and its neighbors at various positions relative to that particle at various times. Clear evidence of a double vortex field at a later time was obtained.

(G) Divergence.

The hydrodynamic model required the solution of the Navier-Stokes equation for the given initial conditions. An analytical solution is only possible at large times. Dimensional arguments at large times showed that the velocity autocorrelation function decays as the reciprocal time to the dimensionality of the system divided by two. Hence, the correlations do not decay exponentially as previously expected and, in fact, in two dimensions decay slowly enough so that the diffusion coefficient diverges. A computer study of the detailed behaviour of the autocorrelation function in two dimensions at long times confirmed not only the power law decay of reciprocal time but also the proportionality constant of that decay.[46] In three dimensions the systems were not quite large enough to observe the asymptotic decay, because boundary interference effects spoiled that observation. The sound as well as the vortex waves constructively interfere at periodic boundaries; an effect which could be quantitatively accounted for. With still larger systems, however, these effects can be sufficiently delayed so as to observe the power law decay in three dimensions as well.

For shorter times, the Navier-Stokes equation must be solved numerically and the numerical solution can be compared both to the vortex field and the velocity autocorrelation. The comparison worked well except at very short times; less than the time corresponding to a few collisions. For such short times, the hydrodynamic model is just not applicable. It is necessary to consider the detailed molecular structure of the fluid. To be sure, by the use of generalized hydrodynamics, which uses time and space dependent transport coefficients in the Navier-Stokes equation, it is possible to apply hydrodynamics at a shorter time scale, but still not at a level of a time between collisions. In fact, the remarkable aspect of these calculations is that the usual hydrodynamics is already valid at such a short time and distance scale, namely at the time corresponding to the order of 10 collisions per particle and a distance scale of the order of a few molecular diameters. Previously it was thought that the molecular and hydrodynamic time scales were far separated. The fact that they are intimately mixed is the real novelty of the molecular dynamics results. The other is the applicability of the hydrodynamic model to the description of the correlation. The model itself has been studied long ago by Stokes.

The three dimensional behavior of the autocorrelation function leads to a convergent diffusion coefficient, however, previous theories have to be modified since they all assumed the validity of a Markovian description at long times. The previous theories can be recovered by projecting out the hydrodynamic part, although, as pointed out above, the separation is not clean cut. Another way to obtain the long time behavior is by means of graph theoretical techniques. These techniques had previously given indication of unusual behavior in the calculation of the density corrections to the Boltzmann low density results. These corrections were shown not to be expressible in a power series in density but involved logarithmic terms in the density. These arose because whole classes of ring diagrams had to be summed in order to get convergent results. For the long time behavior, ring diagrams to all orders must also be summed. These are similar to the cyclical correlations mentioned earlier and the physical analogy to vortex motion is tempting, although this connection has not so far been clearly established.

In two dimensions, because of the divergence, the situation is much more confusing. Neither the Boltzmann equation nor Navier-Stokes hydrodynamics is valid and a new formulation has not yet been clearly found. Even the long time behavior of the volocity autocorrelation function is not known for certain. Although the molecular dynamics calculation established the reciprocal time behavior, which has been confirmed by both the hydrodynamic model and graph theoretical resummation techniques, it is unlikely the correct behavior, in as much as the computer calculations are restricted to small hydrodynamic times and the graph theoretical techniques have similar restrictions. The theoretical hydrodynamic result cannot be believed because it assumes the validity of the Navier-Stokes description which is the very thing that has been proven inapplicable. The realization of this dilema manifests itself in the assumption of the existence of a viscosity. The viscosity and thermal conductivity (at least their kinetic parts) can be shown to diverge also in two dimensions. These transport coefficients have a kinetic as well as a potential part, and whether the potential part diverges is at the moment still an open question.[44] In any case, the divergence of the kinetic part of the viscosity puts the asymptotic behavior of the velocity correlation in doubt. A hand waving, self consistency criterion predicts a velocity autocorrelation divergence as the reciprocal of the product of the time and the square root of the logarithm of the time.[48] Such a behavior would also be consistent with the limiting long time behavior observed by molecular dynamics, since it is numerically impossible to distinguish this behavior from that of reciprocal time.

(H) Brownian Motion

There are some interesting considerations as to what happens to this vortex correlation as the mass and size of the diffusing particle varies. Both in the limit of very light and very massive particles this correlation should disappear. In the Lorentz limit, that is the light particle, the particle scatters among a set of stationary particles and, hence, does not have the momentum necessary to generate a vortex. An interesting question is then whether such a light particle still has a slowly decaying velocity correlation function. Such a slow decay has been found for a perfect gas particle scattered from particles of square-shape in two dimensions.[49]
The origin of this decay is not related at all to the hydrodynamic vortex mode, but arises from trajectories which after many scattering events return the particle to its origin. The long return paths are intimately connected with the geometry and dimensionality of the scatters and it would be interesting to study these correlations for scattering from spheres. Such studies could determine the size above which the light particle is trapped among its neighbors at any given

density. At that size, the diffusion coefficient would vanish because of the high probability of backscattering. The velocity autocorrelation at that point does not necessarily have a long tail. Furthermore, the Enskog theory fails badly because of the neglect of backscattering in that theory.[50] A more appropriate theoretical model would be a percolation model.

For the massive particle, that is the Brownian motion limit, the persistence should be enhanced. However, in the infinite mass or stationary particle limit, it is off-hand hard to see how the vortex mode is established when the particle has no velocity at all. A preliminary study shows that the vortex is created by natural fluctuations leading to a thermally generated sound wave which smashes into the stationary sphere.[51] This sound were manifests itself by repeated successive collisions of the same solvent particle with the Brownian particle. This shows up in an unexpected positive structure in the force autocorrelation function, whose time integral determines the friction constant. This sound wave then leads to a temporary positive pressure on one side of the stationary object that relieves itself also by a vortex flow. Thus, the force autocorrelation function of a finite sized infinite mass particle also has a power law decay in time, just like the velocity autocorrelation function. Only for the infinite size and mass particle does this vortex flow make a vanishing contribution. In that limit, the Stokes hydrodynamic model makes a prediction of the value of the friction constant. The proportionality constant in that prediction depends on the nature of the boundary conditions between the fluid and the Brownian particle. For a perfectly elastic sphere the prediction is that the friction constant is 4π times the viscosity of the medium times the radius of the sphere. Curiously, by molecular dynamics, it was found at densities where the mean free path is small compared to the size of the stationary particle, and, hence, where hydrodynamics should be applicable, that the proportionality constant was 6π; a value obtained only with sticking boundary conditions hydrodynamically. The implication is that boundary conditions are poorly understood from a molecular point of view, and that very likely in this case there must be, on a molecular scale, a boundary layer of solvent spheres around the stationary particle, whose net effect is to allow momentum transverse to the collisions also to be transferred, so as to justify the validity of the sticking boundary conditions.

In this connection, it should be pointed out that, nevertheless, the empirical observations for pure hard spheres are nearly in conformity with the Stokes-Einstein relation that uses 4π.[44] The significance, however, is unclear, since the Stokes-Einstein formula, which relates the diffusion coefficient with the viscosity, is not meant to apply for spheres of the same mass and size as the solvent.

(I) Burnett Coefficients

Although the ordinary transport coefficients were found to exist in three dimensions, the hydrodynamic and graph theoretical models predict divergence of the higher order transport coefficients. These transport coefficients, called Burnett coefficients, are concerned with wave length and gradient dependence. The wave length dependence, characterized by the linear Burnett coefficients, accounts for the dependence of the transport coefficients on the rate at which the external field is applied, while the gradient dependence described by the non-linear Burnett coefficients accounts for the dependence on the amplitude of the external field. The linear Burnett coefficients can be evaluated by higher order autocorrelation functions. The first linear Burnett coefficient for diffusion, for example, can be expressed as the integra of an auto-correlation function of the velocity at four different times Alternatively, the Einstein-like expression for this Burnett coefficier involves the mean of the fourth cumulant of the displacement of a particle. This can be easily programmed for molecular dynamics.
It was indeed shown that this Burnett diffusion in two dimensions, namely for hard disks, does diverge with one higher power in time than the ordinary diffusion coefficient; in agreement with the predictions of the models. The proportionality constant is, however, not in agreement with the prediction. Its density dependence is such that in the low density limit disagreement with the Chapman-Enskog expansion of the Boltzmann equation was found. Although these results have not been completely digested as yet, it is already clear that a power series expansions about the hydrodynamic limit in three dimensions diverges. How to formulate the higher order corrections to the Navier-Stokes equation is, hence, an open question. A clue, still preliminary, as to the reason for these divergences was obtained from a molecular dynamics study of the distribution in displacements. Since all the cumulants of this distribution higher than the first seem to be divergent, it appeared possibly that the distribution would not be Gaussian in the long time limit. Indications are that this is the case.
It appears that the vortex mode restricts the large displacements of particles, because the flow pattern is such that these particles have velocity components that returns them to the origin. What was found, qualitatively consistent with this idea, was that the Gaussian distri-

bution, previous thought valid, was only accurate for small displacements. The distribution had fewer large displacements than predicted by the Gaussian distribution.

(J) Depolarized Light

The trouble so far is that none of these long correlations have been directly observed experimentally. Besides the computer generated evidence, the only real manifestation of this vortex mode appears to be an indirect measurement, namely an enhanced diffusion coefficient at intermediate densities from what one would have predicted if the vortex correlations were left out.[52] The lack of experimental evidence is not surprising in view of the smallness of the effect. The effect is largest for the diffusion coefficient; about 40% enhancement at intermediate densities. Indeed, the few accurate diffusion experiments at these densities and high temperatures are considerable more accurately predicted when interpreted by a hard sphere theory which includes the long tail correction, than by the Enskog theory, which ignores these correlations.

It would be desirable to obtain more evidence, and, therefore, an interpretation of a light scattering experiment was undertaken.[53] The line shape of depolarized light in fluid rare gases can be expressed as the Fourier transform of an autocorrelation function, namely, the induced dipole autocorrelation function. The idea was that the line shape close to the center of the band was a sensitive probe of the long time correlations. Although the line shape near the center presents some experimental difficulties, it could potentially be obtained by special techniques. Furthermore, it must be realized that perhaps not all autocorrelation functions have long tails, so that the one appropriate to light scattering may not be the one to study.

Although this study was interesting in its own right, it failed to give evidence of long tails. The autocorrelation function calculation could explain the band shape, in so far as it is known experimentally, remarkably well. There is, so far, no experimental evidence of a long tail because the central shape has not yet been determined, and the computer results were largely negative that there would be one. It appears that only velocity dependent correlations have long tails, while only spacially dependent correlations, as the induced dipole autocorrelation, may behave normally in that respect. Much further work is required to understand both when long lasting correlations occur and what their consequences are, if they do.

REFERENCES

(1) B.J. Alder and T.E. Wainwright: Transport Properties in Statistical Mechanics. New York: Interscience (1958).

(2) R. Kubo: J. Phys. Soc. Japan 12, 570 (1957).

(3) W.T. Ashurst and W.G. Hoover, Phys. Rev. Letters 31, 206 (1973).

(4) W.W. Wood and F.R. Parker: J. Chem. Phys. 27, 720 (1957).

(5) J.M. Hammersley: Monte Carlo Methods. J. Wiley, New York (1955).

(6) F.H. Ree and W.G. Hoover: J. Chem. Phys. 46, 4181 (1967).

(7) B.J. Alder, S.P. Frankel, and V.A. Lewinson: J. Chem. Phys. 23, 417 (1955).

(8) N. Metropolis, Rosenbluth, Rosenbluth, Teller and Teller: J. Chem. Phys. 21, 1087 (1953).

(9) W. Feller: Introduction to Probability Theory and its Application. New York (1966).

(10) J.G. Sinai: Theory of Dynamical Systems, Lecture Notes Series 23, Aarhus University (1970).

(11) B.J. Alder and T.E. Wainwright: J. Chem. Phys. 31, 459 (1959).

(12) J.B. Gibson, A.N. Goland, M. Milgram and G.H. Vineyard: Phys. Rev. 120, 1229 (1960).

(13) R.H. Miller: J. Comp. Phys. 8, 449 (1971).

(14) B. Widom: J. Chem. Phys. 39, 2808 (1963).

(15) M.H. Kalos: Phys. Rev. A2, 250 (1970).

(16) J.H. Weiner and A. Askar: J. Chem. Phys. 54, 3534 (1971).

(17) Physics of Simple Liquids, Ed. H. N. V. Temperley, J.S. Rowlinson and G.S. Rushbrooke: North Holland, Amsterdam (1968).

(18) B.J. Alder and T.E. Wainwright: J. Chem. Phys. 33, 1439 (1960).

(19) B.J. Alder and T.E. Wainwright: Phys. Rev. 127, 395 (1962).

(20) J.E. Mayer and W.W. Wood: J. Chem. Phys. 42, 4268 (1965).

(21) W.G. Hoover and B.J. Alder: J. Chem. Phys. 46, 686 (1967).

(22) Z.W. Salsburg and W.W. Wood: J. Chem. Phys. 37, 798 (1962).

(23) W.G. Hoover and F.H. Ree: J. Chem. Phys. 49, 3609 (1968).

(24) B.J. Alder, W.R. Gardner, T.K. Hoffer, N.E. Phillips and D.A. Young: Phys. Rev. Letters 21, 732 (1968).

(25) B.J. Alder, W.G. Hoover and T.E. Wainwright: Phys. Rev. Letters 11, 5181 (1963).

(26) M. Kac, G.E. Uhlenbeck and P.C. Hemmer: J. Math. Phys. 4, 216 (1963).

(27) R.W. Zwanzing: J. Chem. Phys. 22, 1420 (1954).

(28) B.J. Alder, D. Young and M. Mark: J. Chem. Phys. 56, 3013 (1971).

(29) B.J. Alder: J. Chem. Phys. 40, 2724 (1964).

(30) B.J. Alder, W.E. Alley and M. Rigby: Physica (in press).

(31) S.H. Sung, D. Chander and B.J. Alder (to be published).

(32) S. Brush, H. Sahlin and E. Teller: J. Chem. Phys. 45, 2102 (1966).

(33) J.P. Hanson, B. Jancovici and D. Schiff: Phys. Rev. Letters 29, 991 (1972)

(34) J.P. Hanson and E.L. Pollock: Phys. Rev. A5, 2651 (1972).

(35) B.L. Holian, W.D. Gwinn, A.C. Luntz and B.J. Alder: J. Chem. Phys. 59, 5444 (1973).

(36) B.J. Alder, D.A. Young, M.R. Mansigh and Z.W. Salsburg: J Comp. Phys. 7, 361 (1971).

(37) W.G. Hoover: J. Chem. Phys. 49, 1981 (1968).

(38) B.J. Alder, B.P. Carter and D.A. Young: Phys. Rev. 183, 831 (1969).

(39) B.J. Alder, W. Hoover and D.A. Young: J. Chem. Phys. 49, 3688 (1968).

(40) R. Grover and B.J. Alder: J. Phys. Chem. Solid. To be published.

(41) B.J. Alder, M. Rigby, A. Sapse and C. Hecht: J. Chem. Phys. 52, 3665 (1970).

(42) T. Einwhohner and B.J. Alder: J. Chem. Phys. 49, 1458 (1968).

(43) J. Dymond and B.J. Alder: J. Chem. Phys. 48, 343 (1968).

(44) B.J. Alder, D.M. Gass and T.E. Wainwright: J. Chem. Phys. 53, 3813 (1970).

(45) B.J. Alder and T.E. Wainwright: Phys. Rev. Letters 18, 988 (1967).

(46) B.J. Alder and T.E. Wainwright: Phys. Rev. A1, 18 (1970).

(47) B.J. Alder and T.E. Wainwright: J. Phys. Soc. Japan 26, 267 (1969).

(48) T.E. Wainwright, D.M. Gass and B.J. Alder: Phys. Rev. A4, 233 (1971).

(49) W.W. Wood and F. Lado: J. Comp. Phys. 7, 528 (1971).

(50) B.J. Alder and W.E. Alley (to be published).

(51) B.J. Alder and P.T. Herman: J. Chem. Phys. 56, 987 (1972).

(52) J. Dymond and B.J. Alder: Ber. der Bunsengesell. Phys. Chem. 75, 394 (1971).

(53) B.J. Alder, H.L. Strauss and J.J. Weiss: J. Chem. Phys. 59, 1002 (1973).

YITZHAK FRISHMAN
Weizmann Institute, Rehovoth, Israel

QUARK TRAPPING IN A MODEL FIELD THEORY

CONTENTS

YITZHAK FRISHMAN
WEIZMANN INSTITUTE, REHOVOTH, ISRAEL

QUARK TRAPPING IN A MODEL FIELD THEORY

I. Introduction

Interaction of an electromagnetic potential with a massless fermion field in one space and one time dimension was used recently to study the question of "quark confinement".[1,2] In this model, the photon propagator acquires a pole at $k^2 = \frac{e^2}{\pi}$ (e the electric charge), and the pole at $k^2 = 0$ disappears.[3,4] The electromagnetic current becomes a free field of mass $\frac{e}{\sqrt{\pi}}$.

The authors of Ref. 1 use the model to study the problem of "quark confinement", namely how is it that we see quark structure at short and light like distances in deep inelastic lepton hadron scattering,[5] while quarks do not come out. As the analog for the electromagnetic current they take a scalar density

$$s(x) = :\bar{\psi}(x)\,\psi(x): \tag{1.1}$$

where ψ is the spinor field. Scalar and pseudoscalar densities are non-trivial in this model. They then consider

$$S(x) = i\langle 0|Ts(x)s(0)|0\rangle \tag{1.2}$$

The imaginary part of the Fourier transform of eq. (1.2) is the analog of $e^+e^- \to$ hadrons total cross section. For the analog of structure functions in deep inelastic lepton hadron scattering they take single particle matrix elements instead of the vacuum to vacuum in eq. (1.2), where the single particle states are one boson states.

For the case of a free massless spinor field, $S(x)$ would be

$$S_{(free)}(x) = \frac{-i}{2\pi^2 x^2} \tag{1.3}$$

Namely a $\frac{1}{x^2}$ term coming form a bubble of massless fermions. In the model under consideration, however, the $\frac{1}{x^2}$ term is completely cancelled by an exponential of a massless propagator with the right coefficient (namely an exponential of $\ell n(-x^2)$), and the result becomes

$$S(x) = \frac{im^2}{2\pi^2} \exp \left[-\pi\Delta_F(m^2,x^2) \right] \tag{1.4}$$

where $m^2 = \frac{e^2}{\pi}$ and

$$\Delta_F(m^2,x^2) = \frac{(-i)}{\pi} \int \frac{d^2k}{k^2-m^2+i\varepsilon} e^{ikx} \tag{1.5}$$

The light cone behaviour of S(x) is the same as in the massless free spinor case, since

$$\Delta_F(m^2,x^2) \underset{x^2\to 0}{\longrightarrow} \frac{1}{\pi} \ln(-m^2x^2) \tag{1.6}$$

However, the exact expression (1.4) shows no zero mass structure. Instead, only massive contributions appear. When expanding the exponential in eq. (1.4), one can interpret the term $\left[\Delta_F(m^2,x^2)\right]^n$ as an n-meson intermediate state, each of mass m.

In this note, we solve the model in momentum space, using Feynman diagrams. Previous treatments[1-4] used x-space methods. The photon propagator is given exactly by the geometric series of the second order bubble, with all other diagrams vanishing. Thus all Green's functions with $n \geq 4$ external vector meson lines vanish. This is shown in Sec. II. In Sec. III we show that there is no scattering amplitudes involving on-shell (mass zero) fermion lines and any mass vector lines. We also show that there is no on-shell fermi-fermi scattering. It is worthwhile to note that in this model one can find a gauge in which the fermion two point function is the free one, while higher 2n point functions of fermi fields are not. What is more relevant, also the gauge invariant parts of the higher 2n point functions are not trivial, like for example the part of the four point function which is S(x). In Sec. IV we discuss the contribution to S(x) and show that the diagrams corresponding to the intuitive picture of Bjorken[6], namely where quarks annihilate pairwise into mesons (see Fig. 10), do not contribute in this model. The only contributions are from radiation of vector mesons from the "original" fermi pair (Fig. 9).

In Sec. V we discuss some more problems. In particular, it is not clear how to distinguish between fermi and bose underlying structure in the discussion of Ref. 1, since the result for S(x) could have been obtained by taking for s(x) an exponential of a massive scalar field to start with. The problem is that the notion of spin exists only for fields but not for states in 1+1 dimensions (the only property is that bose operators commute and fermi operators anticommute, but creation and annihilation operators have no spin index).

II. Solution by Feynman Diagrams

We start with the Lagrangian

$$L = \frac{1}{2}\,\overline{\psi} i \overleftrightarrow{\not\partial} \psi - \frac{1}{4}\, F_{\mu\nu}F^{\mu\nu} - e j_\mu A^\mu \tag{2.1}$$

where ψ is the spinor field, A^μ the electromagnetic field, and

$$\left.\begin{aligned} F_{\mu\nu} &= \partial_\mu A_\nu - \partial_\nu A_\mu \\ j_\mu &= :\overline{\psi}(x)\gamma_\mu\psi(x): \\ \overleftrightarrow{\not\partial} &= \gamma^\alpha(\overrightarrow{\partial}_\alpha - \overleftarrow{\partial}_\alpha) \end{aligned}\right\} \tag{2.2}$$

The Lagrangian (2.1) describes the interaction of a spinor with the electromagnetic field. We work in one space - one time dimensions, in which case the coupling constant e (the electric charge) has dimension of mass.

We are going to solve the model in momentum space. Let us start with the photon self energy. Choosing the Feynmann gauge, we have for the photon propagator

$$D^{(0)}_{\mu\nu}(k) = g_{\mu\nu}\,\frac{(-i)}{k^2} \tag{2.3}$$

The fermion propagator is

$$S^{(0)}(k) = \frac{i}{\not k} \tag{2.4}$$

We take $g_{00} = -\,g_{11} = 1$.

The second order photon self-energy diagram (Fig. 1) yields

$$\tilde{\pi}^{(2)}_{\mu\nu}(p) = -\,\frac{e^2}{(2\pi)^2}\int d^2k\ \mathrm{Tr}\left[\gamma_\mu \frac{1}{\not k}\gamma_\nu \frac{1}{\not p+\not k}\right] \tag{2.5}$$

Fig. 1 Second order photon self energy.

However, this quantity is not transverse. This is easy to see since, in 1+1 dimensions

$$\gamma_\alpha \not{k} \gamma^\alpha = 0 \tag{2.6}$$

which implies that $g^{\mu\nu}\tilde{\pi}^{(2)}_{\mu\nu}(p) = 0$, and hence $\tilde{\pi}^{(2)}_{\mu\nu}(p) = \left(g_{\mu\nu} - \frac{2p_\mu p_\nu}{p^2} \right) \tilde{\pi}^{(2)}(p^2)$. Note that if one integrates over angles first, $\tilde{\pi}^{(2)}_{\mu\nu}(p)$ is actually finite. To see this we note that the possibly divergent part is

$$\int d^2k \, \mathrm{Tr}\left[\gamma_\mu \frac{1}{\not{k}} \gamma_\nu \frac{1}{\not{k}} \right] = C g_{\mu\nu}$$

and by taking a trace with respect to μ,ν we get that $C = 0$ (This is true, for the quantity C, both in the infrared and ultraviolet. However, there is no infrared divergence in the original integral eq. (2.5) for $p \neq 0$).

We thus define the second order photon self energy as

$$\pi^{(2)}_{\mu\nu}(k) = - \frac{e^2}{(2\pi)^2} \lim_{M\to\infty} \int d^2k \, \mathrm{Tr}\left[\gamma_\mu \left(\frac{1}{\not{k}} \gamma_\nu \frac{1}{\not{p}+\not{k}} - \frac{1}{\not{k}-M} \gamma_\nu \frac{1}{\not{p}+\not{k}-M} \right) \right] \tag{2.7}$$

A straightforward calculation yields

$$\pi^{(2)}_{\mu\nu}(k) = \frac{ie^2}{\pi} \left(g_{\mu\nu} - \frac{k_\mu k_\nu}{k^2} \right) . \tag{2.8}$$

The summation of bubbles, Fig. 2, yields for the propagator

$$D_{\mu\nu}(k) = \frac{(-i)}{k^2 - \frac{e^2}{\pi}} \left(g_{\mu\nu} - \frac{k_\mu k_\nu}{k^2} \right) + \frac{(-i)}{k^2} k_\mu k_\nu \tag{2.9}$$

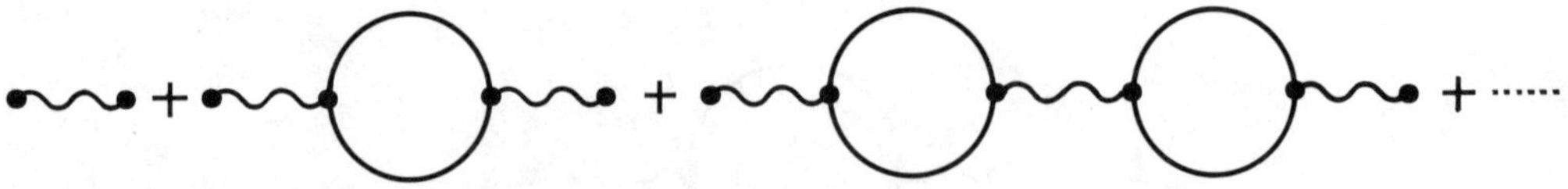

Fig. 2 Bubble summation for the photon propagator.

We thus see that the transverse part represents a vector meson of mass $m = \frac{e}{\sqrt{\pi}}$ (the longitudinal part is, obviously, the same as of $D^{(0)}_{\mu\nu}(k)$).
We are now going to show that eq. (2.9) is, in fact, the exact answer for the propagator, namely that all other diagrams give a vanishing contribution.
We shall do that in several stages. First, we show that diagrams of n external vectors, $n \geq 4$, and one fermion loop, vanish (n odd vanishes trivially). We first take non-exceptional momenta for the external lines (namely, no linear combination is light like), to avoid infrared divergences. Denote by $T^{(n)}_{\alpha_1 \ldots \alpha_n}(k_1 \ldots k_n)$ a one fermion loop with n external vector meson lines of momenta $k_1 \ldots k_n$ and vector indices $\mu_1 \ldots \mu_n$. $T^{(n)}$ is certainly finite for $n \geq 4$. Moreover, unlike $\tilde{\pi}^{(2)}_{\mu\nu}$ of eq. (2.5), it is transverse in any of the vector indices. This is so since the integrand behaves like k^{-n}, and thus even after multiplying by momentum factors one may shift origin of integrations (needed in proving transversality). Note that in order for transversality to hold, we need to add all graphs with a given n, namely all possible permutations of vector lines. This is illustrated in Fig. 3 for the case n = 4. Note that the transversality property holds for massive fermions as well.

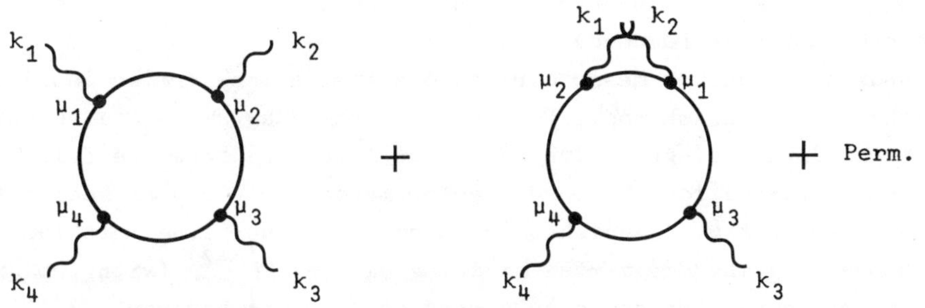

Fig. 3 The diagrams for $T^{(4)}_{\mu_1\mu_2\mu_3\mu_4}(k_1k_2k_3k_4)$.

Another property of the $T^{(n)}$ is the zero trace in any pair of indices. This holds, actually, for any individual diagram that contributes to $T^{(n)}$ (like, for example, the first in Fig. 3 for n = 4), and is a consequence of the zero mass of the fermion. This follows from the fact that the integrand of any such diagram has zero trace. The latter fact is immediate from (particular to 1+1 dimentions),

$$\gamma_\alpha \gamma_{\lambda_1} \cdots \gamma_{\lambda_{2n+1}} \gamma^\alpha = 0 \qquad (2.10)$$

and we always have an odd number of γ-matrices between any two vertices. Eq. (2.10) is a direct consequence of eq. (2.6) and $\gamma_\alpha\gamma_\beta = g_{\alpha\beta}+\varepsilon_{\alpha\beta}\gamma_5$, with $\varepsilon_{\alpha\beta} = -\varepsilon_{\beta\alpha}$, $\varepsilon_{01} = 1$, $\gamma_5 = \gamma_0\gamma_1$.

Take now a $T^{(n)}_{\alpha_1\alpha_2\ldots}(k_1k_2\ldots)$. We have

$$\begin{aligned} T^{(n)}_{00\ldots}(k_1k_2\ldots) &= T^{(n)}_{11\ldots}(k_1k_2) \\ &= \frac{(k_1)_0(k_2)_0}{(k_1)_1(k_2)_1} T^{(n)}_{00\ldots}(k_1k_2\ldots) \end{aligned} \qquad (2.11)$$

where the first step follows from zero trace and the second from conservation.
Since k_1 and k_2 are arbitrary, we conclude

$$T^{(n)}_{\alpha_1\ldots\alpha_n}(k_1 \ldots k_n) \equiv 0 \qquad (2.12)$$

This vanishing must hold also for the exceptional points, by a limiting procedure. Note that this proof breaks down in the case of two external lines, since $\tilde{\pi}^{(2)}_{\mu\nu}$, which is traceless, is not conserved, and $\pi^{(2)}_{\mu\nu}$, which is conserved, is not traceless (since a regulator massive fermion was introduced).
What about vector meson corrections inside the fermion loop? We claim that those vanish too. For consider the diagram in Fig. 4 which is a particular diagram with one inner line. We take the full contribution of Fig. 2 for the inner vector meson. One way of seeing that the diagram in Fig. 4 vanishes is to cut the inner line into two, thereby expressing the diagram as an integral over $T^{(n+2)}$ (with two external momenta equal, which is an exceptional point; however, (2.12) holds everywhere). Another way (which avoids using (2.12) for exceptional momenta) is to choose the bare vector propagator in a particular gauge, which guarantees that the sum of Fig. 2 has only a $g_{\mu\nu}$ part. The required choice is

$$\tilde{D}^{(0)}_{\mu\nu}(k) = \frac{(-i)}{k^2}\left(g_{\mu\nu} - \frac{k_\mu k_\nu}{k^2}\right) + \frac{k_\mu k_\nu}{k^2}\,\frac{(-i)}{k^2 - \frac{e^2}{\pi}}$$

$$= \frac{(-i)}{k^2} g_{\mu\nu} + \frac{e^2}{\pi} \frac{k_\mu k_\nu}{k^2} \frac{(-i)}{k^2\left(k^2 - \frac{e^2}{\pi}\right)} \qquad (2.13)$$

And then we have, instead of the expression on eq. (2.9), only

$$\tilde{D}_{\mu\nu}(k) = \frac{(-i)}{k^2 - \frac{e^2}{\pi}} g_{\mu\nu} \qquad (2.14)$$

The vanishing of the contribution in Fig. 4 then follows from eq. (2.10). The generalization to many inner photons is now trivial.

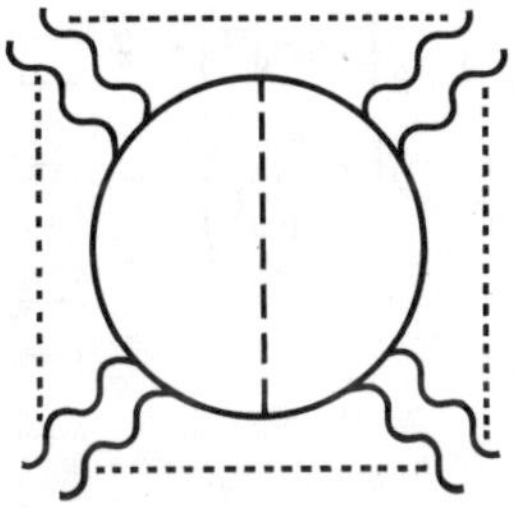

Fig. 4 A contribution of an internal vector meson to an n-external one loop diagram. The dashed line represents the sum of Fig. 2.

With the above results it is now obvious that only the contributions of Fig. 2 remain, and that the exact vector meson propagator has a transverse part

$$d_t(k^2) = \frac{(-i)}{k^2 - \frac{e^2}{\pi}} \qquad (2.15)$$

This shows that the A_μ field is a free field of mass $\frac{e}{\sqrt{\pi}}$, and therefore so is also j_μ . We thus have

$$\left(\Box + \frac{e^2}{\pi}\right) j_\mu = 0 \qquad (2.16)$$

Note also, that in the gauge eq. (2.14), the full fermion propagator is equal to the free one. This is immediate to show by use of eq. (2.10) and the fact that connected n current vacuum expectation values vanish. Thus the fermion propagator remains free. Note that the fermi field is not a free field, since the higher Green's functions, like the 4 point function, are not like the free ones (see section IV). This shows that the Federbush-Johnson theorem[7] does not apply here, since we do not have locality and a positive metric in any one gauge.

III. Scattering Amplitudes

An immediate conclusion from eq. (2.16) is the divergence of the axial current in 1+1 dimensions.[4] The axial current a_μ is

$$a_\mu = :\bar{\psi}\gamma_5\gamma_\mu\psi: = \varepsilon_{\mu\nu}j^\nu \tag{3.1}$$

We have

$$\varepsilon_{\mu\nu}\partial^\nu\left(\partial^\alpha a_\alpha\right) = \varepsilon_{\mu\nu}\varepsilon_{\alpha\beta}\partial^\nu\partial^\alpha j^\beta$$

$$= \left(g_{\mu\beta}g_{\nu\alpha} - g_{\mu\alpha}\,g_{\nu\beta}\right)\partial^\nu\partial^\alpha j^\beta = \Box\, j_\mu = -\frac{e^2}{\pi}\,j_\mu$$

$$= -\frac{e}{\pi}\,\partial^\nu F_{\mu\nu} = \frac{e}{\pi}\,\partial^\nu\left(\varepsilon_{\mu\nu}F\right)$$

where $F = \frac{1}{2}\,\varepsilon^{\alpha\beta}F_{\alpha\beta}$. Thus

$$\partial_\lambda\left(\partial^\alpha a_\alpha - \frac{e}{\pi}\,F\right) = 0$$

and hence

$$\partial^\alpha a_\alpha = \frac{e}{2\pi}\,\varepsilon^{\alpha\beta}F_{\alpha\beta} \tag{3.2}$$

This is an anomaly, since the original Lagrangian is formally invariant under γ_5-transformations.

Another result is that any amplitude, with $n \geq 2$ external vector lines and 2m fermion lines, the latter on their mass shell, is zero. The proof is similar to the one which yields the vanishing of the $T^{(n)}$ (see previous section). When some fermions are off shell this is not true any more, since now the amplitudes are not transverse in the vector indices, and we have Ward identities instead. Thus the amplitude in Fig. 5 is zero when the external two fermions are on shell.

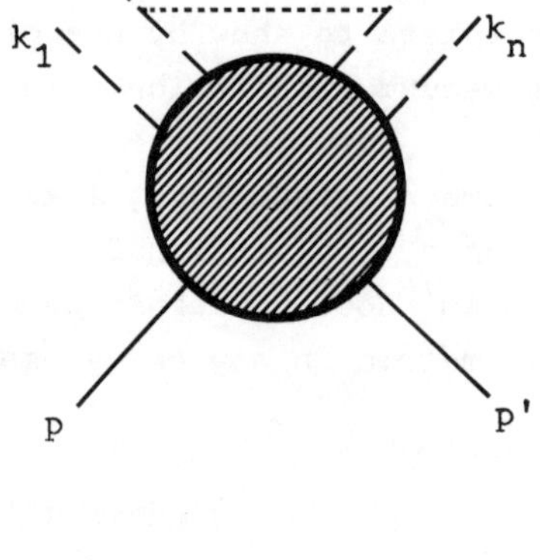

Fig. 5 $n \geq 2$ vector lines and 2 fermion lines Green's function

When considering fermion-fermion scattering, we note that all contributions with at least two photons exchanged vanish. When a photon line is connected to the same fermion line, this is a consequence of eq. (2.10) (we work in the gauge eq. (2.14)). When all photon lines go from one fermion to the other, like in Fig. 6, the vanishing is only for on mass shell fermions, and then it follows from the previous result (vanishing of contribution of Fig. 5 for on shell fermions). Note that for off shell fermions the contribution of Fig. 6 does not vanish, thus giving a non trivial connected 4 point function.

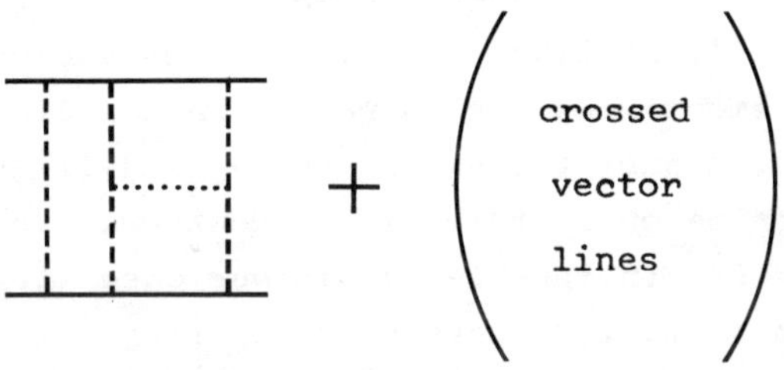

Fig. 6 Contribution to 4 fermion Green's function with $n \geq 2$ inner vector lines.

Coming back to on shell fermion-fermion scattering, the only contribution is then the one vector meson exchange, Fig. 7. The contribution is

$$\left[\bar{u}(p_1')\gamma_\mu u(p_1)\right]\left[\bar{u}(p_2')\gamma^\mu u(p_2)\right] \frac{(-i)e^2}{\left(p_1-p_1'\right) - \frac{e^2}{\pi}} - \left(p_1' \leftrightarrow p_2'\right) \tag{3.3}$$

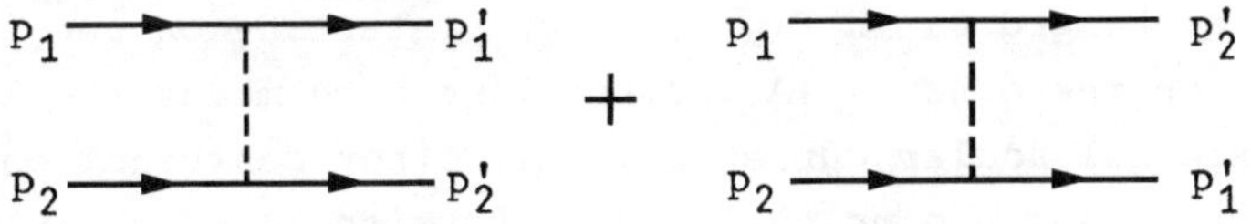

Fig. 7 One vector exchange contribution to fermion-fermion scattering.

Let us consider the first term first. Suppose p_1 is moving in the positive space direction. Then $\not{p}_1\, u(p_1) = 0$ implies $E_1(\gamma_0-\gamma_1)u(p_1)=0$, which means $\gamma_5 u(p_1) = u(p_1)$. Consider now the matrix $u(p_1)\bar{u}(p_2')$. It can be expanded in the complete set 1, γ_5, γ_λ. However, the γ_λ part does not contribute, due to eq. (2.6). From $\gamma_5 u(p_1) = u(p_1)$ it follows that $u(p_1)\bar{u}(p_2')$ has a $(1+\gamma_5)$ combination and not $(1-\gamma_5)$. Thus $\gamma_5\, u(p_2') = -u(p_2')$, and in order for $\bar{u}(p_2')\gamma^\mu\, u(p_2)$ not to vanish we need $\gamma_5\, u(p_2) = -u(p_2)$. Similarly, $\gamma_5 u(p_1') = u(p_1')$. Thus, for the first term not to vanish, p_1 and p_2 are opposite and so are p_1' and p_2', with p_1' parallel to p_1 and p_2' parallel to p_2. In such a case the second term in eq. (3.3) vanishes. Also, $(p_1-p_1')^2 = 0$, since p_1 and p_1' are parallel and light like. Eq. (3.3) thus yields the number $2i\pi$. In the S matrix element we would therefore have a $\delta(p_1-p_1')\delta(p_2-p_2')$ from the disconnected part (minus $p_1' \leftrightarrow p_2'$), and $2i\pi\, \delta^{(2)}(p_1+p_2-p_1'-p_2')$ from the one vector meson contribution. However, since $\delta(p_1-p_1')\delta(p_2-p_2') = \delta^{(2)}(p_1+p_2-p_1'-p_2')$ in our case (trivial to see in center of momentum frame), we have no scattering ($(1+2i\pi)$ can be absorbed in the phase and normalization of free states).

We thus conclude that there is no scattering in this model. However, off shell fermion Green's functions are non trivial. The latter fact will be used in the next section.

IV. Diagrams for Scalar Density Propagator

In order to calculate non trivial quantities in this model, we turn to scalar and pseudoscalar densities. S(x) of eq. (1.2) is here the analog of the usual total cross section for $e^+e^- \to$ hadrons, with s(x) of eq. (1.1) the current instead of the vector current. Diagramtically, the only contributions to S(x) are when the vector lines are exchanged as in Fig. 8, namely starting on one fermion line and ending on the other. (A fermion line here means the line between the two external scalar currents). All other contributions, like vector lines starting and ending on the same fermion line, or interactions between the vector meson lines inside the bubble, are zero (by our previous results). One has to sum the infinite series of Fig. 8 to get the result of Ref. 1. Note that in terms of intermediate states, the structure is like in Fig. 9, namely each diagram has a fermion-antifermion pair and a certain number of vector mesons. It is the infinite summation that wipes out the fermion-antifermion contribution that corresponds to the result of Ref. 1.

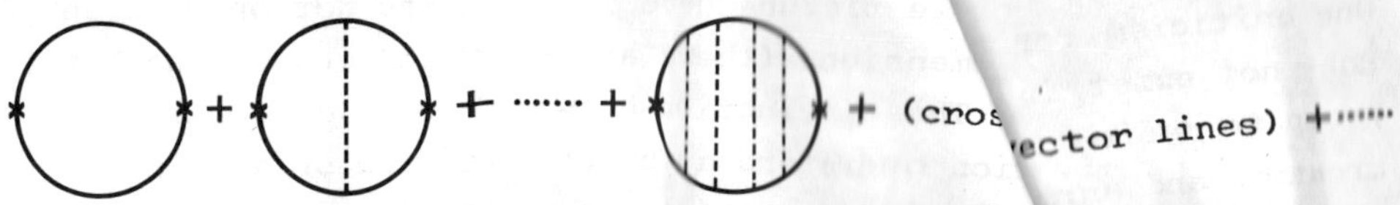

Fig 8 Contributions to S of eq. (1.

Note, that the original intuitive picture of Bjorken in which it is interaction between many pairs of fermions, that binds them to mesons, does not hold here. Namely, the diagrams corresponding to that intuitive picture, like that in Fig. 10, are zero here (a direct result of eq. (2.10), or again from the fact that one has now transversality for any vector index and tracelessness in any pair of vector indices). The screening takes effect only through diagrams like Fig. 9.

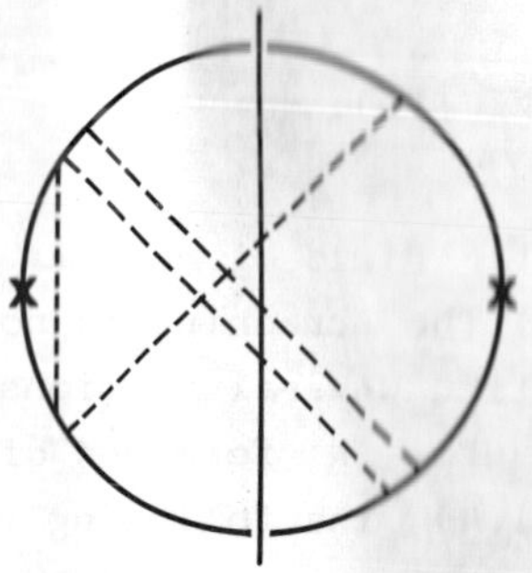

Fig. 9 A contribution to discontinuity of one diagram from the series in Fig. 8.

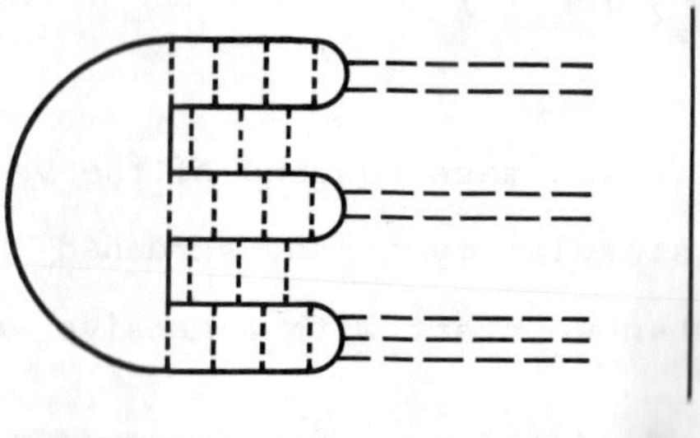

Fig. 10 An intermediate state with "bound" fermion pairs (zero contribution to eq. (1.2)).

V. Problems and Comment

One criticism for the picture here is that the notion of spin does not exist in 1+1 sions (there are no rotations of the little group to define spin he distinction between boson or fermion creation and annihi operators is by commutation or anti-commutation relations in indices exist only for fields as the transformation law und he homogeneous Lorentz group, but not for states. In usual deep in tic lepton hadron scattering, spin $\frac{1}{2}$ constituents lead to transver cross sections while spin 0 constituents lead to longitudinal cr s sections. There is no such analogy here. In Ref. 1 it is sugges d that one looks at the structure of $S(x)$, and then one discovers that the $1/x^2$ term coming from the spinor pair gets cancelled by an exponential of a massless scalar propagator, and one remains with an exponential of a massive scalar propagator. However, exactly the same result for $S(x)$ as in Ref. 1 would be obtained if one chooses instead of s of eq. (1.1) the density

$$\tilde{s}(x) = \frac{m}{\pi\sqrt{2}} : \exp \sqrt{2\pi}\, \phi(x): \qquad (5.1)$$

where $\phi(x)$ is a free scalar field of mass m.

Other problems: (a) The screening is not there when more than one fermion is present, like several fermions whose numbers are separately conserved.[8,9] For N fermions of charges $e_i (i=1,2,\ldots,N)$, we have instead of eq. (1.4), the following for fermion j,

$$S_j(x) = i\langle 0|Ts_j(x)s_j(0)|0\rangle$$

$$= \frac{iM^2}{2\pi^2}\left(-M^2x^2\right)^{-\left[1-\frac{e_j^2}{\sum_{i=1}^{N} e_i^2}\right]} \exp\left[-\frac{e_j^2}{M^2}\Delta_F(M^2,x^2)\right] \qquad (5.2)$$

with $M^2 = \frac{1}{\pi}\sum_{i=1}^{N} e_i^2$, the mass squared of the vector meson in this case. Thus the zero mass singularity is not screened.

(b) No screening when we start with a massive vector meson instead of a massless one.[9]

In this case, instead of the expression eq. (1.4) we have

$$S(x,\mu_o) = \frac{i\mu^2}{2\pi^2} \left(-\mu^2 x^2\right)^{-\frac{\mu_o^2}{\mu^2}} \exp\left[-\frac{e^2}{\mu^2} \Delta_F \left(\mu^2, x^2\right)\right] \qquad (5.3)$$

where μ_o is the bare vector meson mass and μ the final one, with $\mu^2 = \mu_o^2 + \frac{e^2}{\pi}$.

Finally, when one starts with massive fermions, the discussion is much more complex. The model has not been solved. The second order bubble now gives.

$$\pi_{\mu\nu}^{(2)} \left(k, m^2\right) = \frac{ie^2}{\pi} \left(g_{\mu\nu} k^2 - k_\mu k_\nu\right) f^{(2)}(k^2, m^2)$$

$$f^{(2)}\left(k^2, m^2\right) = \int_0^1 d\alpha \frac{\alpha(1-\alpha)}{\alpha(1-\alpha)k^2 - m^2} = \frac{1}{k^2} \left\{ 1 + \frac{2m^2}{k^2} \left(1 - \frac{4m^2}{k^2}\right)^{-1/2} \times \ln \left[\frac{\sqrt{1 - \frac{4m^2}{k^2}} + 1}{\sqrt{1 - \frac{4m^2}{k^2}} - 1} \right] \right\} \qquad (5.4)$$

Thus f has no pole at $k^2 = 0$ and consequently the vector meson remains massless. To get a singularity one now needs a bound state at zero mass.[3] Note that the imaginary part, for $k^2 > 4m^2$, is singular as $\left(1 - \frac{4m^2}{k^2}\right)^{-1/2}$ when $k^2 \to 4m^2$. This is due to the "smallness" of space in one space dimension.

Acknowledgements

This work started in Summer 1973 when the author visited SLAC. It is a pleasure to acknowledge fruitful discussions with J.D. Bjorken and other members of the theory group there. The final version was written during a visit to the Max Planck Institute at Munich in July 1974. The hospitality of Prof. H.P. Dür and other members there is acknowledged.

References

1. A. Casher, J. Kogut and L. Susskind, Phys. Rev. Letters 31, 792 (1973); Cornell University preprint CLNS-251 (August 1973).

2. J. Lowenstein and J. Swieca, Ann. Phys. (N.Y.) 68, 172 (1971).

3. J. Schwinger, Phys. Rev. 128, 2425 (1962).

4. L.S. Brown, Nuovo Cim. 29, 617 (1963);
C. Sommerfield, Ann. Phys. (N.Y.) 26, 1 (1963);
B. Zumino, Acta Physica Austriaca, Supplementum II (Schladming School), p. 212 (1965).

5. For a review of this field see: Y. Frishman, "Light Cone and Short Distances", Weizmann Institute preprint, to appear in Physics Reports.

6. J.D. Bjorken, informal remarks during the "Deep Inelastic Phenomenology" session of the New York Academy of Science Conference, "Recent Advances in High Energy Physics", March 15-17, 1973 (Barbizon Hotel, New York City).

7. P.G. Federbush and K.A. Johnson, Phys. Rev. 120, 1926 (1960).

8. T.M. Yan, Private communication (Aug. 1973).

9. G. Segre and W.I. Weisberger, Preprint UPR 0027T (March 1974).

Lecture Notes in Physics

Bisher erschienen / Already published

Vol. 1: J. C. Erdmann, Wärmeleitung in Kristallen, theoretische Grundlagen und fortgeschrittene experimentelle Methoden. 1969. DM 20,–

Vol. 2: K. Hepp, Théorie de la renormalisation. 1969. DM 18,–

Vol. 3: A. Martin, Scattering Theory: Unitarity, Analyticity and Crossing. 1969. DM 16,–

Vol. 4: G. Ludwig, Deutung des Begriffs physikalische Theorie und axiomatische Grundlegung der Hilbertraumstruktur der Quantenmechanik durch Hauptsätze des Messens. 1970. DM 28,–

Vol. 5: M. Schaaf, The Reduction of the Product of Two Irreducible Unitary Representations of the Proper Orthochronous Quantummechanical Poincaré Group. 1970. DM 16,–

Vol. 6: Group Representations in Mathematics and Physics. Edited by V. Bargmann. 1970. DM 24,–

Vol. 7: R. Balescu, J. L. Lebowitz, I. Prigogine, P. Résibois, Z. W. Salsburg, Lectures in Statistical Physics. 1971. DM 18,–

Vol. 8: Proceedings of the Second International Conference on Numerical Methods in Fluid Dynamics. Edited by M. Holt. 1971. DM 28,–

Vol. 9: D. W. Robinson, The Thermodynamic Pressure in Quantum Statistical Mechanics. 1971. DM 16,–

Vol. 10: J. M. Stewart, Non-Equilibrium Relativistic Kinetic Theory. 1971. DM 16,–

Vol. 11: O. Steinmann, Perturbation Expansions in Axiomatic Field Theory. 1971. DM 16,–

Vol. 12: Statistical Models and Turbulence. Edited by M. Rosenblatt and C. Van Atta. 1972. DM 28,–

Vol. 13: M. Ryan, Hamiltonian Cosmology. 1972. DM 18,–

Vol. 14: Methods of Local and Global Differential Geometry in General Relativity. Edited by D. Farnsworth, J. Fink, J. Porter and A. Thompson. 1972. DM 18,–

Vol. 15: M. Fierz, Vorlesungen zur Entwicklungsgeschichte der Mechanik. 1972. DM 16,–

Vol. 16: H.-O. Georgii, Phasenübergang 1. Art bei Gittergasmodellen. 1972. DM 18,–

Vol. 17: Strong Interaction Physics. Edited by W. Rühl and A. Vancura. 1973. DM 28,–

Vol. 18: Proceedings of the Third International Conference on Numerical Methods in Fluid Mechanics, Vol. I. Edited by H. Cabannes and R. Temam. 1973. DM 18,–

Vol. 19: Proceedings of the Third International Conference on Numerical Methods in Fluid Mechanics, Vol. II. Edited by H. Cabannes and R. Temam. 1973. DM 26,–

Vol. 20: Statistical Mechanics and Mathematical Problems. Edited by A. Lenard. 1973. DM 22,–

Vol. 21: Optimization and Stability Problems in Continuum Mechanics. Edited by P. K. C. Wang. 1973. DM 16,–

Vol. 22: Proceedings of the Europhysics Study Conference on Intermediate Processes in Nuclear Reactions. Edited by N. Cindro, P. Kulišić and Th. Mayer-Kuckuk. 1973. DM 26,–

Vol. 23: Nuclear Structure Physics. Proceedings of the Minerva Symposium on Physics. Edited by U. Smilansky, I. Talmi, and H. A. Weidenmüller. 1973. DM 26,–

Vol. 24: R. F. Snipes, Statistical Mechanical Theory of the Electrolytic Transport of Non-electrolytes. 1973. DM 20,–

Vol. 25: Constructive Quantum Field Theory. The 1973 "Ettore Majorana" International School of Mathematical Physics. Edited by G. Velo and A. Wightman. 1973. DM 26,–

Vol. 26: A. Hubert, Theorie der Domänenwände in geordneten Medien. 1974. DM 28,–

Vol. 27: R. Kh. Zeytounian, Notes sur les Ecoulements Rotationnels de Fluides Parfaits. 1974. DM 28,–

Vol. 28: Lectures in Statistical Physics. Edited by W. C. Schieve and J. S. Turner. 1974. DM 24,–

Vol. 29: Foundations of Quantum Mechanics and Ordered Linear Spaces. Advanced Study Institute Held in Marburg 1973. Edited by A. Hartkämper and H. Neumann. 1974. DM 26,–

Vol. 30: Polarization Nuclear Physics. Proceedings of a Meeting held at Ebermannstadt October 1–5, 1973. Edited by D. Fick. 1974. DM 24,–

Vol. 31: Transport Phenomena. Sitges International School of Statistical Mechanics, June 1974. Edited by G. Kirczenow and J. Marro. 1974. DM 39,–

Vol. 32: Particles, Quantum Fields and Statistical Mechanics. Proceedings of the 1973 Summer Institute in Theoretical Physics held at the Centro de Investigacion y de Estudios Avanzados del IPN – Mexico City. Edited by M. Alexanian and A. Zepeda. 1975. DM 18,–

This series aims to report new developments in physical research and teaching – quickly, informally and at a high level. The type of material considered for publication includes:

1. Preliminary drafts of original papers and monographs
2. Lectures on a new field, or presenting a new angle on a classical field
3. Seminar work-outs
4. Reports of meetings, provided they are
   a) of exceptional interest and
   b) devoted to a single topic.

Texts which are out of print but still in demand may also be considered if they fall within these categories.

The timeliness of a manuscript is more important than its form, which may be unfinished or tentative. Thus, in some instances, proofs may be merely outlined and results presented which have been or will later be published elsewhere. If possible, a subject index should be included. Publication of Lecture Notes is intended as a service to the international physical community, in that a commercial publisher, Springer-Verlag, can offer a wider distribution to documents which would otherwise have a restricted readership. Once published and copyrighted, they can be documented in the scientific literature.

Manuscripts

Manuscripts should comprise not less than 100 pages.

They are reproduced by a photographic process and therefore must be typed with extreme care. Symbols not on the typewriter should be inserted by hand in indelible black ink. Corrections to the typescript should be made by pasting the amended text over the old one, or by obliterating errors with white correcting fluid. Authors receive 50 free copies and are free to use the material in other publications. The typescript is reduced slightly in size during reproduction; best results will not be obtained unless the text on any one page is kept within the overall limit of 18 x 26.5 cm (7 x 10½ inches). The publishers will be pleased to supply on request special stationery with the typing area outlined.

Manuscripts in English, German or French should be sent to Dr. W. Beiglböck, 69 Heidelberg/Germany, Institut für Angewandte Mathematik, Im Neuenheimerfeld 5, or directly to Springer-Verlag Heidelberg.

Springer-Verlag, D-1000 Berlin 33, Heidelberger Platz 3
Springer-Verlag, D-6900 Heidelberg 1, Neuenheimer Landstraße 28–30
Springer-Verlag, 175 Fifth Avenue, New York, NY 10010/USA

ISBN 3-540-07022-2
ISBN 0-387-07022-2